www.wadsworth.com

www.wadsworth.com is the World Wide Web site for Thomson Wadsworth and is your direct source to dozens of online resources.

At *www.wadsworth.com* you can find out about supplements, demonstration software, and student resources. You can also send email to many of our authors and preview new publications and exciting new technologies.

www.wadsworth.com
Changing the way the world learns®

Current Perspectives
Readings from InfoTrac® College Edition

Introductory Sociology Research Updates

Australia • Canada • Mexico • Singapore • Spain
United Kingdom • United States

Current Perspectives: Readings from InfoTrac® College Edition
for Introductory Sociology Research Updates

Acquisitions Editor: *Bob Jucha*
Assistant Editor: *Elise Smith*
Editorial Assistant: *Christina Cha*
Technology Project Manager: *Dee Dee Zobian*
Marketing Manager: *Wendy Gordon*
Marketing Assistant: *Annabelle Yang*
Marketing Communications Manager: *Linda Yip*
Project Manager, Editorial Production: *Brenda Ginty*
Creative Director: *Robert Hugel*

Print Buyer: *Karen Hunt*
Production Service: *Rozi Harris, Interactive Composition Corporation*
Permissions Editor: *Joohee Lee*
Cover Designer: *Larry Didona*
Cover Image: *Photolibrary.com/Photonica*
Cover and Text Printer: *Thomson West*
Compositor: *Interactive Composition Corporation*

© 2006 Thomson Wadsworth, a part of The Thomson Corporation. Thomson, the Star logo, and Wadsworth are trademarks used herein under license.

ALL RIGHTS RESERVED. No part of this work covered by the copyright hereon may be reproduced or used in any form or by any means—graphic, electronic, or mechanical, including photocopying, recording, taping, Web distribution, information storage and retrieval systems, or in any other manner—without the written permission of the publisher.

Printed in the United States of America
1 2 3 4 5 6 7 09 08 07 06 05

For more information about our products, contact us at:
**Thomson Learning Academic Resource Center
1-800-423-0563**

For permission to use material from this text or product, submit a request online at
http://www.thomsonrights.com.
Any additional questions about permissions can be submitted by email to
thomsonrights@thomson.com.

Library of Congress Control Number: 2004117884

ISBN 0-495-00764-1

Thomson Higher Education
10 Davis Drive
Belmont, CA 94002-3098
USA

Asia (including India)
Thomson Learning
5 Shenton Way
#01-01 UIC Building
Singapore 068808

Australia/New Zealand
Thomson Learning Australia
102 Dodds Street
Southbank, Victoria 3006
Australia

Canada
Thomson Nelson
1120 Birchmount Road
Toronto, Ontario M1K 5G4
Canada

UK/Europe/Middle East/Africa
Thomson Learning
High Holborn House
50-51 Bedford Row
London WC1R 4LR
United Kingdom

Latin America
Thomson Learning
Seneca, 53
Colonia Polanco
11560 Mexico
D.F. Mexico

Spain (including Portugal)
Thomson Paraninfo
Calle Magallanes, 25
28015 Madrid, Spain

Contents

Preface   vii

1. Sociological Perspectives   1
   Sociological Imaginings and Imagining Sociology: Bodies, Auto/biographies and Other Mysteries
   DAVID MORGAN, *Sociology*

2. Sociological Research Methods   18
   Monitoring Costs and Tolerance Levels for Classroom Cheating
   GARY GALLES, PHILIP E. GRAVES, ROBERT L. SEXTON, and SURREY M. WALTON, *The American Journal of Economics and Sociology*

3. Culture   24
   School Culture: Exploring the Hidden Curriculum
   DAVID J. WREN, *Adolescence*

4. Socialization   28
   Racial Socialization and Racial Identity: Can They Promote Resiliency for African American Adolescents?
   DAVID B. MILLER, *Adolescence*

5. Society, Social Structure, and Interaction   37
   Longitudinal Perspective: Adverse Childhood Events, Substance Use, and Labor Force Participation Among Homeless Adults
   TAMMY W. TAM, CHERYL ZLOTNICK, and MARJORIE J. ROBERTSON, *American Journal of Drug and Alcohol Abuse*

6. Groups and Organizations   54
   My Culture, My Self (Cultural Differences in Japan and the United States)
   BRUCE BOWER, *Science News*

7. Deviance and Crime   59
   The Effect of Video Games on Feelings of Aggression
   DEREK SCOTT, *The Journal of Psychology*

8. Global Stratification   70
   Value-Adding Information: Virtual Conferencing, a Telecommunication Pathway to the Future
   JOAN EDGECUMBE, *Nursing Administration Quarterly*

9. Social Class in the United States   79
   America's Emphasis on Welfare: Is It Children's Welfare or Corporate Welfare?
   SALLY RAPHEL, *Journal of Child and Adolescent Psychiatric Nursing*

10. Race and Ethnicity 84
    The Origins and Demise of the Concept of Race
    CHARLES HIRSCHMAN, *Population and Development Review*

11. Sex and Gender 116
    Women, Disability, and Sport and Physical Fitness Activity: The Intersection of Gender and Disability Dynamics
    ELAINE M. BLINDE and SARAH G. MCCALLISTER, *Research Quarterly for Exercise and Sport*

12. Aging and Inequality Based on Age 134
    Older People and the Enterprise Society: Age and Self-Employment Propensities, Work, Employment and Society
    FIONA WILSON, JAMES CURRAN, and ROBERT A. BLACKBURN, *International Small Business Journal*

13. The Economy and Work in Global Perspective 138
    Long-Term Unemployment Among Young People: The Risk of Social Exclusion
    THOMAS KIESELBACH, *American Journal of Community Psychology*

14. Politics and Government in Global Perspective 151
    Mandatory Co-Operation: Because Water Knows No Political Boundaries, Water-Sharing Nations Must Work Together
    JENNIFER PEDERSEN and EVAN D.G. FRASER, *Alternatives Journal*

15. Families and Intimate Relationships 154
    Is Posthumous Semen Retrieval Ethically Permissible?
    R. D. ORDD and M. SIEGLER, *Journal of Medical Ethics*

16. Education 164
    A Choice Between Public and Private Schools: What Next for School Vouchers?
    PAUL E. PETERSON, *Spectrum: The Journal of State Government*

17. Religion 172
    And the Wisdom to Know the Difference? Freedom, Control and the Sociology of Religion
    EILEEN BARKER, *Sociology of Religion*

18. Heath, Health Care, and Disability 196
    Have You Checked Your Pension Plan's Funding Level Lately?
    ANGELO CALVELLO, *Healthcare Financial Management*

19. Population and Urbanization 200
    Unwise Use: Gale Norton's New Environmentalism
    DAVID HELVARG, *The Progressive*

20. Collective Behavior, Social Movements, and Social Change 207
    A Retrospective on the Civil Rights Movement: Political and Intellectual Landmarks
    ALDON D. MORRIS, *Annual Review of Sociology*

InfoMarks: Make Your Mark 230

Preface

The social world in which we exist is marked by continuous change. This perpetual revolution is a result of the limitless aspects of the human experience that constitute our day-to-day existence. Because the human social experience is intriguing and ever changing—from the interpersonal to the cultural to the global—we are encouraged to research and learn about the changes that occur around us. Whether we are talking about perspectives, research methods, or collective behavior, there is much to explore, question, and value. In the many journals, books, and studies written and generated in the field of sociology, sociologists have attempted to compile and update the most recent information pertinent in their field; yet, they have only just begun to scratch the surface of available resources.

In *Current Perspectives: Readings from InfoTrac® College Edition: Introductory Sociology Research Updates,* we invite you, the instructor and student, to learn from and utilize some of the tools that we have available to keep you current on emerging research in the field of sociology. This reader has been carefully designed to include recent *InfoTrac* articles on each aspect of the core Introduction to Sociology curriculum, covering topic areas such as global stratification, aging, health care, and social movements, among others. These articles have been carefully selected with the Introductory Sociology curriculum in mind. This reader is intended for the Introductory Sociology course in general, and more specifically for courses that use Andersen/Taylor's *Sociology: Understanding a Diverse Society,* Brym/Lie's *Sociology: Your Compass for a New World,* Ferrante's *Sociology: A Global Perspective,* Kornblum's *Sociology in a Changing World,* Tischler's *Introduction to Sociology,* and Kendall's *Sociology in Our Times.*

The articles have been chosen from periodicals across sociology, and other related disciplines, to provide a wide range of discussion materials that can be used to supplement any Introductory Sociology course curriculum. For instance, Derek Scott's article "The Effect of Video Games on Feelings of Aggression" from *The Journal of Psychology* will help introduce you to the discussion of whether video games contribute to deviance and crime in contemporary societies. David J. Wren's "School Culture: Exploring the Hidden Curriculum" from *Adolescence* offers insight into the influence of teenage culture within the high school educational setting. These articles also can provide additional readings to supplement the boxed section of most Wadsworth Sociology texts, often titled "Doing Sociological Research" or "Sociology in Global Perspective." These current articles help expose both students and instructors to the ever-changing sociological world, and provide material for discussions and writing assignments on foundational sociological concepts. Most importantly, the articles will help students begin to understand how sociologists write, think, and research.

In addition to providing students and instructors with recent articles to utilize in a classroom setting, this reader gives students and instructors the tools to begin their own original research. Using *InfoTrac College Edition's* new *InfoMark* capabilities, students and instructors can create their own *InfoTrac* Virtual Readers compiled of *InfoTrac* articles. With minimal effort, this resource allows you to search, compile, and create a list of articles that can be formatted into an electronic reader. Detailed instructions are included at the back of this reader on how to find and fully use this new resource at http://www.infotrac-college.com.

InfoTrac College Edition is a complete online research and learning center with more than 10 million full-text articles from nearly 5,000 scholarly and popular periodicals (including *The New York Times*). These articles are updated daily and date back to 1980. Included are some major sociological journals, such as *American Journal of Sociology, Sociological Quarterly,* and *Society*. This fully searchable library is available from any computer with Internet access. In addition to this, *InfoTrac* includes *InfoWrite*. This will help you to develop your research writing skills, including such tasks as choosing topics, crediting sources, developing thesis statements, grammar and word usage, using quotations, composing introductions and conclusions, writing drafts and revisions. Moreover, *InfoTrac* now has a Critical Thinking resource center. That is, it provides instructions about how to determine whether or not a particular argument ought to be believed. This includes how to distinguish facts from opinions, primary from secondary sources, evaluating information, and recognizing deceptive arguments and stereotypes.

We invite you to use *Current Perspectives: Readings from InfoTrac College Edition: Introductory Sociology Research Updates* as a stepping-stone to complement and expand the discussion and research components of your Introductory Sociology course. Using this resource as a supplement to your main Introductory Sociology textbook, combined with *InfoTrac College Edition,* we hope you will enhance your understanding and knowledge of current research in the field of sociology.

As always, we welcome comments and suggestions from you as you use this reader in your Introductory Sociology courses. Please email us with any feedback.

Elise Smith
Assistant Editor, Sociology
Thomson Higher Education
elise.smith@thomson.com

1

Sociological Perspectives

Sociological Imaginings and Imagining Sociology: Bodies, Auto/biographies and Other Mysteries

David Morgan

This paper seeks to explore sociology as an imaginative pursuit. After a brief reconsideration of Mills's notion of 'the sociological imagination' I examine three areas illustrating the various imaginations within the discipline: the work of Robert K. Merton; ethnomethodology; and the diversities of feminist scholarship. Two particular case studies are explored: the sociology of the body and the use of auto/biographical studies in sociology. I conclude with some suggestions for the encouragement of imaginative thought within the discipline.

I begin my address with two stories. The first concerns one of my early job interviews in or around 1962. In the course of the interview, one of the interviewers asked: 'Which British sociologist do you regularly read for pleasure?' Apart from the fact that the notion of reading sociologists for pleasure struck me then (although not now) as slightly whimsical, the number of British sociologists that could even make the short-list seemed very few indeed. American, yes; I could have gone on for quite a while about

C. Wright Mills. But British? Of course, the only possible answer at that time would have been the first President of this Association (1955–57), Morris Ginsberg, but then I had only read, with some bewilderment, his short volume entitled *Sociology* (1934). I did not get the job.

Some years later, now securely employed at Manchester University, I was discussing an essay with a third-year student. In the course of the discussion she expressed her enthusiasm for Max Weber stating that he 'kept her awake at night'. Wisely, I resisted the rather obvious flippant response to the effect that Weber usually had the opposite effect on most students. There were serious issues at stake, and indeed, I shared her disturbed fascination with this particular sociologist. This was something deeper and more complex than a simple intellectual challenge, and I suppose that the main justification for preserving something called 'the sociological canon' is that writers with this kind of capacity to disturb are rare enough in any generation, however dead, white or male they may be.

My subject, therefore, is to do with sociology as an imaginative pursuit, a practice that is capable of delighting and disturbing. I want to try to explore some of the different strands of this imaginative pursuit. I want to look back, selectively, over what sociology has achieved to gain intimations of what sociology might continue to be. I shall attempt to apply these reflections particularly to two recent areas of study. The first, the subject of this conference, deals with the body, while the second deals with the growing interest in auto/biographical practices in sociology and many other social sciences and areas of the humanities. In my conclusion, I shall attempt to suggest how we might cultivate this imagination and why, at this particularly moment in our history, it might be especially important to think about these matters.

It will be noted that I use the word 'mysteries' in my title. This may not be entirely accidental. Of the many meanings of this word provided by the Oxford English Dictionary several of them revolve around a kind of distinction between the sacred and the secular. Sacred mysteries are a source of awe and reverence, objects of worship. Secular mysteries are puzzles to be solved. It may be thought that the history of sociological thought is of a shift from the sacred to the secular, but I hope to be able to show that matters may not be so simple as that. Game and Metcalfe's use of the idea of 'wonder' has some affinities with my concerns here (Game and Metcalfe 1996).

SOCIOLOGICAL IMAGININGS

The title of this lecture, of course, has more than just an echo of C. Wright Mills, whose writings, when I was beginning sociology, certainly provided more than their fair share of stimulation and pleasure. The pluralities in the title may reflect a nod in the direction of postmodern sensibilities, a recognition of diversities. The reference to 'imagining sociology', yet again serves as a reminder of the various understandings of 'reflexivity' that have developed and been elaborated following Gouldner's discussion of the term in *The Coming Crisis of Western Sociology* (1971). The key idea here is of a sociology constantly reflecting critically and creatively upon its own practices, a form, as Gouldner reminds us, of work ethic. I would also want to convey a sense that there is an imaginary element in sociology, a striving for something that is always just out of reach and that it is the process of reaching out rather than the completed act of grasping that is important.

Mills's *The Sociological Imagination* will, no doubt, continue to inspire students and researchers although it is likely that this inspiration will be increasingly qualified. This is not simply a function of the ungendered, yet covertly masculine, character of the text, something which is most likely to strike the modern reader turning to it for the first time. The very title might suggest a problem, pointing to 'The' sociological imagination with its unambiguous singularity. However, his famous definition could encompass a range of imaginations: 'The sociological imagination enables us to grasp history and biography and the relations between them within society. That is its task and its promise' (1970:12).

His equally celebrated dismissal of 'grand theorists' and 'abstracted empiricists', with the implied denial of the label 'imagination' to the supposed practitioners operating at these two extremes, seems to me now to be ungenerous. The thoughtful reading of a Parsons or a Lazarsfeld may provide some routes towards this imaginative grasping (the key word is probably 'enables') especially when it is realised that sociological enquiry is carried out in a context of relatively open debate and criticism. One notes, for example, how the volumes of research into *The American Soldier* (much criticised by Mills) inspired Merton's fruitful middle-range theorising around the idea of reference groups just as, today, their re-reading may provide insights into the role of the military in the construction of masculinities (Stouffer et al. 1949a, 1949b).

Yet this is not the only description of the key sociological task. Towards the end of the book he writes: 'What social science is properly about is the human variety, which consists of all the social worlds in which men have lived, are living and might live' (Mills 1970:147). This sense of plurality is not, it would appear, applied to the various overlapping communities of social scientists or to the varieties of ways in which these multiple social worlds might be grasped. Yet again, his strictures on the straitjackets of methodology and his endorsement of a sense of playfulness in his appendix (Mills 1970:233) suggest that these considerations might not, after all, be too far away.

What perhaps is at stake is what is to be understood by the word 'imagination' in this title. It is clear that the links between biography, history and society can be achieved mechanically with relatively little imaginative understanding. Mills's phrase 'the cheerful robot', for example, might be seen in terms of the links between these three elements although the description now seems little more than a cliché, more the product of the author's construction rather than of the late modern society. It is also true that imagination may be found in writers who do not necessarily or overtly place Mills's project to the forefront of their activities.

I want to convey a sense of the diversity of imaginations, and their limitations, in considering a range of authors and approaches which might appear to have little to do with each other. In the first place, consider Robert K. Merton, an author who does not appear in The Sociological Imagination, despite the considerable support and encouragement he is said to have given to Mills (Horovitz 1983). Whatever criticisms that may be leveled at, for example, Merton's functionalism or the over-formally structured character of his model of deviancy he could not be said to lack imagination. His discussion of functionalism, for example, still contains much that is potentially shocking as, for example, where he considers the possible societal functions of political rackets or the political machine (Merton 1957:71–82). In addition, consider the range of ideas and themes which he either originated or developed further: the self-fulfilling prophecy, the Matthew principle, reference groups, the paradox of unintended consequences, manifest and latent functions (Clark, Modgil and Modgil 1990). Above all, perhaps, it is the tone which expresses the imagination over and above any particular content: detached, ironic but continuously curious and questioning, a tone which Stinchcombe notes in arguing for the importance of an aesthetic approach in Merton's sociology (Stinchcombe 1990: 90–1).

This ironical stance would almost certainly not appeal to my second example of sociological imaginings, namely the work of Garfinkel and ethnomethodologists. The apparent claim on the part of social analysts to probe beneath the surface of everyday life, to disclose the real or hidden meanings and motivations on the part of social actors would be, to the ethnomethodologists, a claim without warrant. While claiming to study everyday life, conventional sociology presents models of that life which more readily reflect the life-worlds of sociologists than everyday life itself.

If this were the sole contribution of ethnomethodology, the call to look critically at the very practices through which sociologists and other analysts claim to describe and account for the social world the end result would certainly justify the title of social imagination since it involved a radical refocusing of perspective and redefinition of the sociological task. Yet there are other more interesting reasons why the ethnomethodological stance should attract our attention. Ethnomethodologists sometimes talk about the 'awesomeness' of social phenomena; Garfinkel referred to 'the strangeness of an obstinately familiar world' (Garfinkel 1967:38). Even if we were to have reservations about some of the strands and themes of the ethnomethodological project, this recognition of 'strangeness' or 'awesomeness' must strike a welcome chord. Such a recognition does not imply a relatively passive or worshipful stance towards the strangeness of everyday social life. Rather, it involves a collaborative willingness on the part of the observer to see the world in a new way, a disciplined and even rigorous refocusing. That such a perspective could, and probably did, become routinised does not detract from the imaginative thrust of the ethnomethodological project.

My third example is much broader than either of the other two and much less easy to characterise in a few simple sentences. I am referring to the contribution of feminist scholarship, recognising the diversity that is covered by this label and, perhaps controversially, including here the responses on the part of men such as Bob Connell, to the broad feminist critique. Here I want to concentrate upon the further, radical refocusing that is demanded when we begin to understand the gendered nature of social phenomena and of the means that we have for the understanding and interpreting the social. What is particularly striking here is the range of practices and perspectives that are included within the feminist critique. While much of the work has focused upon areas of life where one might expect to find a critical exploration of the persistence of gender inequalities and exploitations such as family, sexual divisions of labour and educational opportunities, these areas have been augmented by a wide-ranging critique that comes right back to the practices of sociology itself. Here too, the focus is not simply upon the theories that have been elaborated within a malestream sociology but upon the recognition that sociologists, like everyone else, occupy multiple lives which impact upon each other and depend upon a range of significant others, such as partners, secretaries, cleaners and so forth, in order to bring forth their sociological output.

Equally impressive is the depth as well as the range implied by feminist scholarship. By this I mean the willingness not to stop simply at exposing patterns of exploitation or inequality (within, say, families or organisations) but to ask a range of questions which look at the very context within which these practices are taking place. Thus, taking an example from family studies, the analysis does not simply stop at the cataloguing or mapping of inequalities within domestic divisions of labour or bringing to the surface some of the unwelcome facts of sexual and domestic violence. It goes further in looking at the very category of family itself, the consequences of understandings focused upon the idea of family and the role of sociologists in the construction of these categories of understanding. In the case of work and organisations again, it is not simply the recognition of patterns of structured inequality and sexual harassment. It is also looking at the very idea of organisation and the way in which these ideas are located within the framework of masculinist rationalities. The increasing use of the word 'gendering' in this and other contexts, implying a complex and collaborative process rather than a thing, is an indication of these continuous processes of reformulation.

Indeed, it would be possible to go further and see the range of issues considered within the feminist scholarship and the depth of particular explorations as criss-crossing each other. Thus questions of violence, not exclusively the result of feminist scholarship but significantly influenced by it, move away from the more obvious considerations of criminality and armed combat to consider violence much closer to home, in families, workplaces, schools and sports. At the same time the more familiar areas of criminality and war are rendered somewhat unfamiliar through their overt gendering on the part of the researcher. Similarly, questions of sexuality move away from paradigmatic models of heterosexual penetration to wider explorations of the erotics of everyday life, the recentring of desire within routine social practices.

Such examples could be multiplied, but I hope that enough has been said to justify my location of feminist and feminist-inspired scholarship within my understanding of sociological imaginings. Once again, the world has been rendered strange and the familiar has been shown to be unfamiliar. Further, these critical understandings do not simply apply to the world 'out there' but reflect back upon the very practices of sociology itself.

I have considered a particular theorist, Robert K. Merton, a particular school or approach, ethnomethodology, and a very wide-ranging political, social and intellectual movement in feminism. Doubtless, there are other examples I could have chosen and some old favourites have been missed.[1] In terms of individuals mention might have been made of Goffman or Hughes, in terms of topic reference might have been made to the significance of a global, more thoroughgoing comparative perspective (as reflected, for example, in debates about orientalism or Islam) or the developments of queer theory. What all these, and many more, have at their best, is a sense of rendering mysterious without mystifying. In all cases, the taken-for-granted world is seen as being in someways strange, through the adoption of a new perspective or a willingness to stand up close or move away.

A key part of my understanding of sociological imaginings is the phrase 'without mystifying'. It may be argued that all the approaches mentioned here—and many others that have not been mentioned—may be prone to their inbuilt mystifying tendencies. One might, indeed, see examples of the routinisation of charisma present in these sociological renewals. In some cases, the routinisation might be reflected in the emphasis on technique or method rather than the process of discovery. In other cases, the routinisation might be in terms of over-elaborate theorising where we are called upon to admire the complexity or awesomeness of the models themselves rather than the social phenomena which they are attempting to capture. In yet other cases, the routinisations might take the form of ever more esoteric or specialized languages, the province of an inward-looking cult rather than a creative school. This, perhaps, is a third set of meanings associated with the word 'mystery'. All these routinisations tend towards mystification and the response of the uninitiated observer might be more one of exasperation than of awe. However, sociology should know enough about these tendencies in many areas of social life—politics, artistic movements, religion—to have the potential to recognise and explore them within the practice of sociology itself.

It may be felt that this emphasis on rendering mysterious represents something of a departure from the radical approach of C. Wright Mills and all those who emphasise changing the world as well as, or instead of, understanding it. The answer to this is partly that to see the world as mysterious does not preclude attempts to change it. Indeed, by showing the world as something other than the taken-for-granted solidarities that most readily or most often present themselves is to show a world that is open to change and challenge. For example, the Marxist analysis of the fetishism of commodities renders the commodity somewhat mysterious by showing it to be something other than the face that it presents to the world but itself exists within a wider critique of the institutions of capitalism that sustain and within which these fetishisms are reproduced (Geras 1972). Similar observations may be made about the feminist explorations of gender and gendering or, indeed, about the way in which, as Rustin argues, ethnomethodology shows the 'arbitrary and indeterminate core' within supposed rationalities (Rustin 1993:177).

EMBODYING SOCIOLOGY

My first extended example is the subject of this conference: the growing sociological, indeed more generally scholarly, interest in the body. There is perhaps more than an element of the traditional mystery story here. In a country house a maid is doing her rounds, early in the morning. She discovers a body in the library and with considerable resourcefulness she goes off to summon the authorities. When they return, however, there is no sign of a body; only books, old and new, scattered about the library. The body apparently, solid and decidedly present, seems to disappear in a collection of texts or noisy discourses. As Game and Metcalfe note: 'Scores of disembodied and dispassionate studies paradoxically urge the significance of bodies and emotions' (Game and Metcalfe 1996:5). There are all kinds of people who might have a motive for removing the body from the library. There are those, presumably a diminishing number, who might consider that the subject matter is not all that nice and does nothing for the reputation of the discipline. There are those who are worried about essentialism or biological reductionism and some of the disturbing thoughts in these directions that might be aroused by the sight of a body, dead or alive. And there are those modern descendants of Burke and Hare who would wish to spirit off the body to their private libraries for secret dissection.

All this to state the obvious, namely that the body is a contested site within the field of scholarly discourse (let alone the wider world outside) and that its place within sociology is still by no means stable or secure. Critics might observe that those who claim to be talking about the body often seem to end in talking about something else: gender, power, governmentality, the self and so on. There would seem to remain a gulf between our numerous, undoubtedly real, individual experiences of pain and desire, confrontations with our reflections in the mirror, senses of delight or discomfort on the one hand and the increasing numbers of discussions of the body and embodiment in sociological texts on the other. There remains something oddly elusive and illusive about something so solid and so intimately linked with our sense of being in the world (Turner 1984:7–8).

Nevertheless, the central achievement of the recent developments in the sociology of the body has been to render mysterious this basic feature of our everyday experience. Of course, sociology is a relatively late arrival in the business of rendering the body mysterious. The inner workings of the human, and other, bodies have frequently been described as mysteries and the explorations of the circulation of blood, the workings of the nervous system and of the processes of reproduction have frequently been likened to the discoveries of new worlds by intrepid travellers. Possibly the use of the word mystery was originally used to keep these explorations within a religious framework and to avert any accusations of blasphemy or attempting to play God. Later, as I have suggested, the term 'mystery' might take on a more secular character referring to puzzles to be solved but still retaining a sense of awe and wonderment not simply at the bodily processes exposed to public gaze but also at the processes of their discovery.

To what extent do sociological investigations of the body retain this sense of mystery, not simply in terms of problems to be solved but also as something mysterious in the older sense of the word? Quite simply, it might be maintained, it is in the recognition of a social dimension to the way in which we understand, experience, control and relate to our bodies as individuals and as members of various collectives. As Laqueur writes: 'Instead of being the consequence of increased scientific knowledge, new ways of interpreting the body were rather . . . new ways of constituting social realities . . . Serious talk about sexuality is inevitably about society' (Laqueur 1987:4).

Examples here might include the debates and elaborations following Parsons's discussion of the sick-role, Foucauldian and post-Foucauldian explorations of the constitution of the body through numerous gazes and Goffman's links between everyday interactional practices and sense of the sacredness surrounding the body. Perhaps a particularly striking example would be Gallagher and Laqueur's exploration of the way in which ideology intervened in the perception of the female body and her reproductive system. A nice example of social anthropological irony would be Miner's exploration of body ritual among Nacirema, placing the proximate acts of teeth-cleaning and other morning ablutions in a wider context of cultural meaning (Miner 1956).

In these, and numerous other examples, the mystery lies in calling into question our routine sense of the everyday and taken for granted, somehow that much more solid when presented in an embodied form, and doing this by linking these routine events, observations or experiences to wider cultural forms or processes. The mystery lies in the simultaneous sense of recognition and a sense of strangeness. Unhappily and probably inevitably, this sense of strangeness can frequently disappear in the more assertive claims about social constructionism and the specialised languages and discourses that become elaborated around these claims. It seems all too easy to slide from a sense of wonder at the connectedness of bodily processes to culture and history to something which sounds like a claim that the body and bodily processes are 'simply' or 'solely' social constructions.

One way of preserving the mysteries is to recognise that it is not simply a question of sociology informing understandings of the body and embodied processes but also one of sociology being enriched by a more embodied understanding. The embodied understanding can inform routine sociological debates and practices. This is obvious, perhaps, in some areas such as the sociologies of gender and of health and illness. It is perhaps a little less obvious in discussions of stratification or organisations. To take the former example, the embodiment of class has been both recognised and unrecognised in sociological discourse. It is recognised, in that the embodied character has always been present in sociological discussions although unrecognised in that these presences remain to be highlighted by the reader rather than strongly present in the texts themselves. The key terms of Marxist analysis—alienation, exploitation, surplus value—directly or indirectly take us back to bodies, their use and deployment within capitalist systems and the raw experiences of workers as they confront these systems on a day-to-day basis. Class consciousness is not only frequently symbolised by bodies of workers both sharing and challenging a common fate but is based upon a recognition of the sharing of physical and symbolic space by embodied workers. In more popular iconography and language, of course, the embodiment of class is the more obvious and striking: in the bloated representations of the capitalists (in more recent times to re-appear as fat cats), the fear of the great unwashed or the body trapped in the mechanically repetitive rhythms of mass production.

These rather traditional representations or embodied understandings of class can be supplemented in modern times by more complex representations of class and status systems. It is not simply that the more complex divisions and hierarchies of modern society make use of and are represented in bodily terms as embodied life styles that interact with social and economic status and other divisions such as age, ethnicity and gender. It is also that embodiment and different modes of being in a body become relatively autonomous systems of status differentiation in their own right. Here one only has to refer to the numerous discourses around health and body management and the ways in which these themes are woven around notions of morally responsible individuals. There can, therefore, be little doubt that embodied ways of thinking can enrich our understanding and the ways in which we think about some of the more abiding concerns of sociology around social divisions.

WRITING SOCIOLOGICAL LIVES

Another area of social enquiry which has received considerable attention in recent years has been around the uses of auto/biography. I here use the term, following Liz Stanley, not simply as shorthand representation of autobiography and/or biography but also in recognition of the inter-dependence of the two enterprises (Stanley 1992). In writing another's life we also write or rewrite our own lives; in writing about ourselves we also construct ourselves as somebody different from the person who routinely and unproblematically inhabits and moves through social space and time.

As with the study of the body, this is frequently an interdisciplinary enterprise and one which involves, within sociology itself, numerous areas of sociological practice. In other words it is not simply a separate area of sociological enquiry or interest but an approach that can inform many different sub-areas as well as the process of sociological enquiry itself. Part of the potentially mysterious character of auto/biographical understandings is the extension of these auto/biographical understandings to a much wide range of practices including the construction of curricula vitae, 'Desert Island Discs' and medical records. The construction of lives is not simply the business of trained or paid specialists but the collaborative work of all of us. Further, these developing notions of auto/biographical practices serve to remind us that these practices, contrary to what might be normally understood, while focusing upon individuals are never purely individual.

The biographer of C. Wright Mills has noted how few social science biographies actually exist (Horowitz 1983:1). However, this is not to say that auto/biographical practices do not exist within social science writings. For example, in a recent issue of *Body and Society*, Tony Jefferson begins a discussion of 'Hard Men' with two personal memories (Jefferson 1998). In developing the theme let me make use of three published examples of tragic and/or painful events which authors have used in the context of developing a sociological (or, in one case, a social historical) argument.

My first example comes from Etzioni's controversial argument outlined in *The Spirit of Community* (1995:145). He describes, in few words, how his wife was injured and subsequently died after a serious car accident one winter and how the neighbours rallied round and supported him in the weeks that followed. He concludes this brief discussion with these words: 'But basically people help one another and sustain the spirit of community because they sense it is the right thing to do' (Etzioni 1995:145). This is one, but probably the most striking, example taken from personal experience in his elaboration of the principles of communitarianism.

My second example comes from a social historian, John Gillis (1996:ix–xi). In the Prologue to his *A World of Their Own Making,* Gillis describes how he, and some members of his family, heard about the death of his son, Ben, in a flying accident in Kenya while waiting for his telephone call on Christmas Day. In the course of this moving account, Gillis describes the complex and contradictory feelings that he had about family, his own and families in general, and about the way in which myths and rituals were involved in these complexities. Christmas, they realised, could never be the same after his death and gradually new rituals, around a vegetarian meal, were developed.

My final example comes from a text which has been bracketed with Mills's *The Sociological Imagination:* Charles Lemert's *Social Things* (1997:ix). Here, the author describes the development of a friendly, if casual, relationship with the owner of a local delicatessen in Brooklyn. One day, Lemert learned that this friendly shopkeeper had been murdered; he was to discover that the murdered man had come with his family from Yemen in order to develop a new life in the United States. Lemert uses this experience in a variety of ways but initially, at least, to highlight the contrast between the abstract mysteries of 'society' and the realities revealed in numerous events and experiences, not necessarily all as dramatic as the one described.

A variety of points may be made about these illustrations. In the first place, none of them is strictly necessary for the arguments being presented; they could all be removed without loss. However, it is clear that they do not simply function as human illustrations to more abstract arguments. What we can see here is the weaving of social science or pedagogical lives with personal lives. In the case of Gillis, this is closest to what might be described as an epiphany, a significant turning point within a life. He writes: 'I never thought I would write a book about myth and ritual. I did not seek out the subject: it was something that found me' (Gillis 1996:ix). While the account is not strictly necessary for the reader exploring the differences between 'the families we live by' and 'the families we live with', for the author the experience was obviously of considerable significance. We have no way of knowing whether he would have developed these arguments without this experience, but it is clear that he understands it in this way.

It would be tempting to see the other two experiences as being of a different order, their use being more rhetorical in the case of Etzioni and pedagogical in the case of Lemert. However, this might be to suggest that there is something less personal, less autobiographical, within the main bodies of these texts. In fact, both texts are full of illustrations, some based on personal experience with others more removed, yet all of them functioning to advance the particular arguments being presented. While we are encouraged to make distinctions between scholarly activity and 'real life', in practice this is not a distinction that holds up to close examination. Both present a distinct, if different, vision of the world within which distinctions between the public and the private or between the scholarly and the personal become blurred if not entirely obliterated. The passion for social enquiry is evident in all three texts and it is this that provides the unity between the conventional oppositions. All these illustrations may, with full recognition to the complexities and ambiguities that each presents, provide some support for Gouldner's assertion: 'Surely truth, no less than error, must be born of social experiences' (Gouldner, 1971:482).

There is no doubt, a lot more that could be said about these illustrations. However, the point that I should like to emphasise is that, as with the case of the recent writings on the body, auto/biographical practices of this kind are good to think with. They show the interdependence between the realm of ideas and theories and everyday life, not simply (as in the case of these three illustrations) in the more dramatic or epiphanal moments but also in the steady, apparently undramatic, flow of everyday life. In a sense the sociological imagination is brought home; it is not simply theft biographies and theft histories with which we are concerned but also our lives and historical contexts which are woven together in the texts we produce and constitute through our readings.

IMAGINING SOCIOLOGY

I have talked about some of the ways in which we can talk about sociological imaginations, a plurality of ways to understand the links between biography and history. However, it is not just this particular linkage which unifies these imaginations for one might say that all sociology, effective or otherwise, addresses these linkages to some degree or another. What I am trying to explore is that capacity to develop a sense of the mysterious. If the use of that word still carries with it hints of its religious origins, so be it. What is implied here is a set of contrasts: between things as they are and things as they seem: between things seen by one actor and set of actors and their appearance to other sets of actors: and between how things seem or are and how they might be. Different sociologists, coming from different sociological perspectives might give different emphases to one or more of these contrasts but each of them introduces at least a note of mystery into the process of sociological analysis. If words like 'mystery' or 'awe' rest uncomfortably within our understanding of social life, then think of the frisson of excitement that accompanies processes of insight or discovery. This is part of the recognition of the routine overlaps between feelings and reason (one of Ginsberg's key emphases according to Fletcher (1974:16–19)) and a recognition that sociological enquiry is not to be exempted from these overlaps.

Another way of talking about this process is of making a difference. It can be argued that almost all sociology makes some kind of difference even if it is only—and this practice should not be belittled—a question of adding to the stock of sociological research findings. But the kind of difference I am talking about is of another order; it is a question of encouraging readers or listeners to see the social world in a slightly different way, to see as problems topics that had previously been taken for granted.

Hence, the sociology I am imagining is one that has the capacity to address itself to social mysteries. It has to be recognised that much sociology has little of the mysterious about it but this is simply to say that it shares a lot in common with all other areas of disciplined enquiry. The kind of imagined sociology is perhaps like coffee, often more enticing in the promising aroma than in the actual consumption. But just as few of us, I suppose, would wish to eliminate the aroma of coffee simply because it promises more than it can deliver, so too the fact that everyday sociology delivers up few mysterious should not put us off the task altogether. Is there any way in which this imagined sociology could be encouraged? It would run counter to some of the key assumptions of the discipline to believe that this could not be done and that we have to continue to rely upon the appearance of the individual genius. In the examples that I have provided there are perhaps some hints of what might be possible. While, following convention, I have frequently referred to individual names it is important to note that these names were embedded in networks of relationships and influences. This is one of the uses of the new developments in auto/biographical analysis. Thus Merton, it might be argued, not only developed the idea of the role set but experienced and practised it in his scholarly activity (Clark, Modgil and Modgil 1990). The same is certainly true of C. Wright Mills despite his reputation as a loner and a maverick (Horowitz 1983). In the cases of ethnomethodology and feminist scholarship, the collective (in the sense of overlapping circles rather than centrally organised collectivities) nature of the enterprises is much more clear cut. It is always worth reminding ourselves that the strength of sociology and its ability to renew itself lies less in talented individuals but in the development and encouragement of groups, schools and networks of scholarly exchange. One need hardly emphasise the importance of the British Sociological Association in this respect (and, of course, all similar organisations) and particularly the work of the study groups and caucuses.

Another area of scholarly practice which is worth pausing to consider is the actual genres of sociological output. Much of the output in our ever expanding number of journals corresponds to Peter Medawar's description of the scientific paper (1990:228–9): First, there is a section called the 'introduction' in which you merely describe the general field in which your scientific talents are going to be exercised, followed by a section called 'previous work' in which you concede, more or less graciously, that others have dimly groped towards the fundamental truths that you are now about to expound. Then a section on 'methods'—that is OK. Then comes the section called 'results'... You reserve all appraisal of the scientific evidence until the 'discussion' section.

Turner and Turner present a rather similar account (1990:115). Medawar was, of course, describing publications in the biological sciences and he asked whether this conventional structure, often insisted upon by journal editors, had its fraudulent character in that it gravely distorted the actual nature of the process of scientific research and discovery. From my perspective, the importance of this piece is in encouraging us to look critically at the routine ways in which we present our work.

At the other end of the continuum from Medawar's scientific paper we might locate the sociological essay. This owes more to the essay as a literary genre represented by some of its best practitioners such as Orwell in more recent times or the classic works of Montaigne, Bacon, Hazlett and others. Here the key characteristics would seem to be a more personal voice, a less predictable structure and a relative absence of scholarly references. Some of the key figures here would include Hughes, Becker, Goffman, Dorothy Smith and, straying a little from sociology, Max Gluckman. It is also worth noting that much of Ginsberg's writing has more of the essay format about it, straying relatively effortlessly over disciplinary boundaries.

The point here is not so much one of arguing for the essay or belletristic format against the scientific paper. There are strengths in the latter just as there are weaknesses in the former. However, given the relative rarity of good sociological writing in the essay genre it might be worth considering the extent to which and the ways in which it might be possible to encourage this form of writing. However, the point here is more one of looking critically at the ways in which we tell our sociological stories and the extent to which the conventional forms might from time to time, inhibit the sharing of imaginative understandings of social processes.

Finally, in this section, it is worth saying a few words on teaching. Here it might be worth noting that the fraudulent qualities that Medawar found in the scientific paper might be repeated and amplified in many of the text-book versions of the social world. Here, as someone who has contributed to this genre, I would include some of my own ventures or participations in this field. Among the problems with the textbook genre, we might list the construction of topics which become associated with more or less reified entities (work, family, class, leisure, etc.) or the perpetuation of an illusion of completeness. However, more recent ventures into the field might be said, in various ways, to challenge these conventional formats and to focus more upon the encouragement of a sociological understanding, upon process rather than upon findings (for example, Ballard, Gubbay and Middleton 1997; Bauman 1990; Game and Metcalfe 1996; Lemert 1997; Runciman 1998; Stones 1996).

Teaching, of course, is not limited to or by textbooks although it could be argued that some of the more recent examples reflect a wider concern to shift the basis of sociological knowledge from outcomes to process. Probably one of the happier consequences of teaching quality assessments was not simply the encouragement of a greater critical awareness of the process of teaching but also a recognition that teaching, like research, was essentially a collective undertaking and responsibility. One of the hopes is that the increasing degrees of surveillance and monitoring of academic and scholarly practices, especially in relation to teaching, will not inhibit the development of more imaginative understandings of what it is to think sociologically.

BY WAY OF A CONCLUSION

My concerns in this lecture have been with sociological imaginings, seeing these in the plural as representing a greater, and more creative, diversity, than was perhaps suggested in Mills's *Sociological Imagination*. It would, for example, seek to find imaginations in some of the works, whether of grand theory or abstracted empiricism, subjected to Mills's strictures. I have attempted to argue that sociology at its best, has the capacity to render mysterious without mystifying and I sought to provide illustrations of this both from the traditions which have informed and stimulated my thinking and some of the more recent developments in the field around the sociology of the body and the uses of auto/biography.

The qualification 'without mystifying' is important. Prophets can easily become priests and sociology is not exempt from the processes of routinisation which have been observed in religious and political life. The excitements of new ways of viewing the world may readily become tracts of private property surrounded by tangled thickets of unfamiliar language as comforting to the insiders as it is a source of irritation to the outsider. John MacKenzie's complaints about some of the writings of what he describes as 'discourse theorists' may serve as an illustration of this kind of process and the complaints which are frequently made about it (MacKenzie 1995:39).

Nevertheless, even here a strand of sociological imagining may be deployed in order to understand and frame the familiar claims and counterclaims about jargon and esoteric cults and practices within the social sciences and elsewhere. Some very familiar sociological notions may be brought to bear to understand the processes and tendencies to which we are all prone from time to time. These would include the routinisation of charisma or the 'nothing fails like success' phenomena, the constant dialectic between insiders and outsiders and the important recognition that vested interests in ideas might be at least as important as more material interests. Such a sociological understanding of the processes of mystification or the tendencies of intellectual excitement to fall back into routines might also remind us of the possible inevitability of such processes even in the face of such awareness.

However much these routinisations may be with us, it is essential to remind ourselves constantly of the mysteries and the diverse imaginations. This may be particularly important in the light of this quotation from the abbreviated version of the Dearing Report, where it talks of the importance of sustaining 'a culture which demands disciplined thinking, encourages curiosity, challenges existing ideas and generates new ones' (National Committee of Inquiry into Higher Education 1997:8). Sociologists may see this call as not simply something to consider in relation to their own practices but also as an invitation to ask themselves how their imaginings might contribute to these processes within the wider culture. There are good reasons to fear that these concerns will get obscured in more immediate debates about funding and standards, the latter perhaps being conceived in somewhat narrower terms than might be suggested by this quotation.

The debates following the Dearing Report remind us of the importance and the immediacy of these concerns. In a similar vein, recent reports of the decline of applications for sociology degrees may also be an occasion to think about the ways in which we stimulate sociological imaginings. How far, for example, has sociology come to seem rather dull and conventional compared with some of the more publicised debates around, say genetics or the origins of the universe?

To some extent this lecture is looking forward to next year when we, in common no doubt with scores of other groups and conferences, take stock of where we have reached and speculate on where we are going. In the meantime I hope that we will continue to be awed by the mysteries of social processes and that we, and our students, may sometimes be kept awake at night by our imaginings.

NOTE

1. It has been pointed out to me that most of the authors cited have been American. This is a reflection of my autobiography and of the key influences in my undergraduate and early postgraduate years. British influences were more likely to be particular clusters of studies (especially community studies) or the work of social anthropologists such as Evans-Pritchard, Gluckman and others. Personal influences would be another matter. Over the years these have included Derek Allcorn, Isabel Emmett, Ronnie Frankenberg, Janet Finch, Max Gluckman, Jeff Hearn, Gordon Horobin, John Lee, Nod Miller, J. Clyde Mitchell, Valdo Pons, Sue Scott, Wes Sharrock, Liz Stanley, Sue Wise and Peter Worsley. The usual apologies are due to those who have been excluded—and included.

REFERENCES

Ballard, C., Gubbay, J. and Middleton, C. (eds.). 1997. *The Student's Companion to Sociology*. Oxford: Blackwell.

Bauman, Z. 1990. *Thinking Sociologically*. Oxford: Blackwell.

Clark, J., Modgil, C. and Modgil, J. (eds.). *Robert K. Merton: Consensus and Controversy*. London: Falmer.

Etzioni, A. 1995. *The Spirit of Community*. London: Fontana.

Fletcher, R. (ed.) 1974. *The Science of Society and the Unity of Mankind*. London: Heinemann.

Game, A. and Metcalfe, A. 1996. *Passionate Sociology*. London: Sage.

Garfinkel, H. 1967. *Studies in Ethnomethodology*. Englewood Cliffs: Prentice Hall.

Eras, N. 1972. 'Marx and the Critique of Political Economy', pp. 284–305 in Robin Blackburn (ed.), *Ideology in Social Science*. London: Fontana.

Gillis, J. R. 1996. *A World of Their Own Making: A History of Myth and Ritual in Family Life*. Oxford: Oxford University Press.

Ginsberg, M. 1934. *Sociology*. London: Thornton Butterworth (Home University Library).

Gouldner, A. W. 1971. *The Coming Crisis of Western Sociology*. London: Heinemann.

Horowitz, I. L. 1983. *C. Wright Mills: An American Utopian*. New York: The Free Press.

Jefferson, T. 1998. 'Muscle. "Hard Men" and "Iron"-Mike Tyson: Reflections on Desire, Anxiety and the Embodiment of Masculinity'. *Body and Society* 4:73–98.

Laqueur, T. 1987. 'Orgasm, Generation and the Politics of Reproductive Biology', pp. 1–41 in C. Gallagher and T. Laqueur (eds.), *The Making of the Modern Body*. Berkeley: University of California Press.

Lemert, C. 1997. *Social Things: An Introduction to the Sociological Life*. New York: Rowman & Littlefield.

Mackenzie, J. M. 1995. *Orientalism: History, Theory and the Arts*. Manchester: Manchester University Press.

Medawar, P. 1990. *The Threat and the Glory: Reflections on Science and Scientists*. Oxford: Oxford University Press.

Merton, R. K. 1957, revd and enlarged edn. *Social Theory and Social Structure*. Glencoe, Ill: The Free Press.

Mills, C. W. 1970/1959. *The Sociological Imagination*. Harmondsworth: Penguin.

Miner, H. 1956. 'Body Ritual among the Nacirema'. *American Anthropology* 58:503–7.

National Committee of Inquiry into Higher Education. 1997. Higher Education in the Learning Society (Summary Report). London: HMSO.

Runciman, W. G. 1998. *The Social Animal*. London: Harper Collins.

Rustin, M. 1993. 'Ethnomethodology', pp. 167–94 in D. Morgan and L. Stanley (eds.), *Debates in Sociology*. Manchester: Manchester University Press.

Stanley, L. 1992. *The Auto/Biographical I*. Manchester: Manchester University Press.

Stinchcombe, A. 1990. 'Social Structure in the Work of Robert Merton', pp. 81–96 in J. Clark, C. Modgil and J. Modgil (eds.) *Robert K. Merton: Consensus and Controversy*. London: Falmer.

Stones, R. 1996. *Sociological Reasoning: Towards a Post-Modern Sociology*. Basingstoke: Macmillan.

Stouffer, S. A. et al. 1949a. *The American Solder: Vol 1. Adjustment During Army Life*. Princeton, NJ: Princeton University Press.

Stouffer, S. A. et al. 1949b. *The American Soldier: Vol 2. Combat and Its Aftermath*. Princeton, NJ: Princeton University Press.

Turner, B. S. 1984. *The Body and Society*. Oxford: Blackwell.

Turner, S. P. and Turner, J. H. 1990. *The Impossible Science: An Institutional Analysis of American Sociology*. Newbury Park: Sage.

2

Sociological Research Methods

Monitoring Costs and Tolerance Levels for Classroom Cheating

Gary Galles, Philip E. Graves, Robert L. Sexton, and Surrey M. Walton

The amount of cheating and plagiarism on college campuses has become an increasingly important topic. The usual policy prescription to this type of problem is to step up monitoring efforts. However, this is difficult and costly. An alternative strategy is to substitute tougher and consistent sanctions for cheating and plagiarism.

I. INTRODUCTION

Cheating and plagiarism are endemic at many American colleges and universities (see Collison 1990; Bunn, Caudill, and Gropper 1992), despite numerous rules attempting to reduce such behavior. The central problem, of course, is the cost of stringently monitoring cheating on examinations or term papers.

The role of policing, and its cost, in the setting of what may usefully be thought of as "optimal cheating limits" is an important but often overlooked feature of policies designed to reduce or eliminate cheating behavior. In a world of costly and imperfect enforcement, university policy makers need to seek methods that make compliance cost-effective. This is particularly so, given recent financial cutbacks facing American universities.

Cheating is usually difficult to detect without costly monitoring and investigation costs. Class size is often large and it is difficult to detect students purchasing a paper or cheating on an exam.

One solution is for the university to be less tolerant of cheating by setting tougher sanctions. With more stringent sanctions, a smaller amount of enforcement cost would be necessary to achieve a given reduction in cheating.[1] For example, a school whose ethics code presently states that an F should be given for the individual paper or exam in which plagiarism or cheating occurred might consider giving a course grade of F to a convicted cheater or an even stiffer penalty-expulsion with the cheating incident recorded on the student's permanent record.

II. ENFORCING THE LIMITS

As already noted, without stiff sanctions, many students might choose to cheat at levels that are nonoptimally large from a social perspective. Specifically, individual students will choose levels (equating marginal private benefits and marginal private costs) that are in excess of those that equate marginal social benefits and costs. This is a classic "externality" case in the jargon of economics. Even with noncurved classes there would be a grade inflation effect damaging the meaning of performance at any grade level (e.g., an A would come to mean less). But the externality is particularly evident in the case of "a curved class," in which the cheater might well impose a lower grade on one or more other students.

What is the optimal "cheating" limit? It would be where the marginal social cost and benefits of cheating were equated, from an economist's perspective. (Note that this includes the benefits to the cheater as true benefits, perhaps a philosophical difficulty for some.) Suppose one takes the private benefits from cheating, properly measured (i.e., how much does the cheater really "benefit" from his or her actions?) as being small and views the social damages to other students as being large. In this case, if regulatory compliance were achievable costlessly, the optimal cheating limit would be zero, or something quite near that. The crucial feature of our model, which incorporates the fact that achieving the optimal cheating limit requires monitoring and enforcement, comes from the recognition that the average cheating level will depend on both tolerance of cheating (reflected in the penalties imposed) and on the cost of policing cheating.

Formally, let the average cheating level in a classroom, C, be a function of both the tolerance for cheating, T, and the level of policing, P.[2] This function is given by $C(T, P)$. Over relevant ranges of T and P, it is reasonable to assume that C increases at a decreasing rate with respect to T and decreases at a decreasing rate with respect to P. That is, the rate of cheating would increase at a decreasing rate with tolerance, since in this case the benefits of cheating would be declining in the number of people attempting to cheat (there would be fewer people worth cheating from as the number attempting to cheat rises, since attempting to cheat implies less study effort). Hence, as more attempt to cheat, the return to personal study rises. Letting subscripts represent partial derivatives with respect to the indicated variable, we have

$$C_1 > 0, C_{11} < 0, C_2 < 0, C_{22} > 0$$

The private net benefit realized from the average cheating level is given by the function B(C). Since the purpose of tougher sanctions is to keep students from cheating as much as they otherwise would, it is assumed that $B'(C) > 0$ over the relevant range, with $B''(C) < 0$. The cost of externalities, E(C), is given as a function of average cheating level only, with $E'(C) > 0$ and $E''(C) > 0$. It is assumed that the marginal and average cost of policing is given by the positive constant O. Finally it is assumed that there is some tolerance level, T, below which it is politically impossible to lower the tolerance level further—say jail time for first time offenders.[3]

We are now in a position to express the objective of the cheating limit policy—maximizing net social benefits—as solving for the T, P, and λ, which maximizes

$$(1) \quad Z(T, P, \lambda) = B[C(T, P)] - E[C(T, V)] - \Theta P + \lambda(T - \underline{T})$$

The Kuhn-Tucker solution to this inequality-constrained maximization problem is:

(2) $\delta Z/\delta T = [B'(C) - E'(C)]C_1 + \lambda \leq 0$
(3) $\delta Z/\delta P = [B'(C) - E'(C)]C_2 - \Theta = 0$
(4) $\delta Z/\delta Z = T - \underline{T} \geq 0$
(5) $T, P, \lambda \geq 0$
(6) $\{[B'(C) - E'(C)]C_1 + \lambda\}T = 0$
(7) $\lambda(T - \underline{T}) = 0$

The intuition behind these conditions is clear. Condition (3) calls for an increase in policing until its marginal value, $[B' - E']C_2$, is equal to its marginal cost, Θ. Since $\Theta > 0$ and $C_2 < 0$, it follows from Equation (3) that $B' - E' < 0$, that is, the rate at which cheating increases private benefits must be less than the rate at which it increases negative externalities. This, along with the fact that $C_1 > 0$, means that is strictly positive in Equation (6), hence $T - \underline{T} = 0$ from Equation (7). Intuitively, the relative resource savings from imposing more stringent sanctions as compared to greater policing effort is being frustrated by failure to lower cheating tolerance levels to the political minimum. In this simple case the optimal cheating level is completely independent of the specific nature of the functions B(C) and E(C), with only the amount of policing being affected by the benefits or costs associated with cheating.

III. APPLYING THE MODEL TO CHEATING

The intuition of the preceding model is that there are only two ways to reduce the average level of cheating: lower the tolerance level by imposing tougher sanctions or raise the level of policing. The former is relatively socially costless while the latter is not, since it costs more to put additional proctors, graders, and additional time in the library or on the Internet to hunt down sources on papers. Obviously, the advantage is in substituting a lower tolerance level—a tougher code of ethics for socially costly policing to the fullest extent possible.

This deterrence approach has been recently challenged (see Gneezy and Rustichini 2000). Fines were imposed to penalize parents that were late picking up their children from day care centers. The result was the number of late parents increased. However, this experiment does not seem to apply to the case of cheating at school for several reasons. First, Gneezy and Rustichini's analysis involves the institution of a fine where there was none before—in contrast, students already know that there is a "fine" for cheating. Second, their loss of predictive power from the deterrence effect was set in a commercial setting where instituting a fine changed participant's attitudes from "this is a favor so don't overuse it" to "this is a price so I can buy all I want at the prevailing price (fine)." But increasing cheating penalties would not lead students to the analogous belief that cheating is now more permissible than before. Finally, it is likely that a nonoptimal fine was imposed on day care parents (the fine was $3 if the parent was more than 20 minutes late for pickup), whereas the penalty prior to the fine was the "guilt or conscience effect" for passing their cost on to others. A monetary fine needs to be established that is greater than the guilt or conscience effect. For example, at $20 we would predict fewer late pickups than at $3.

It is possible that some students have cheated through school to get to college or are socialized to believe that everybody cheats, so perhaps they will not respond to tougher cheating sanctions by cheating less often. While this minority of students could be less responsive to more effective anti-cheating policies, it is unlikely that they would not respond at all, given the resources universities make available for those who wish to learn how to do their work properly (writing centers, tutorial programs, mandatory freshman English, and so on). Further, if there are some students who do get to college after it is "too late" to alter their amount of cheating, that is not an argument against imposing more effective cheating approaches, but rather an argument to begin to impose tougher sanctions at an earlier age.

IV. CONCLUSIONS

Once the cost of enforcing cheating is recognized, the tolerance level for cheating should, on efficiency grounds, be set as low as is feasible. The feasible level may well depend on what most faculty and administrators believe is "fair."

Many might find, say, automatic expulsion for the slightest infraction to be a bit too draconian. Yet surely the current climate of public opinion toward purchased papers and cheating on examinations would lead to optimal cheating levels below the prevailing levels. At least over some range, the cost of lowering the amount of cheating can be reduced by substituting tougher sanctions for costly detection efforts. This will not only move us toward more socially desirable levels of cheating, but university administrators will welcome the resource cost savings in times of tight budgets.

Our analysis implies that the efficient tolerance level for cheating will be as low as politically possible. However, we did not investigate what would determine that minimum possible tolerance level. One limiting factor would be the point at which faculty believed penalties were so severe that they would not prosecute cases of cheating—say, expulsion for copying a friend's answer on a homework problem. That is, if the penalties are not viewed as just, there may be a common disrespect for the law and the level of cheating could actually rise. Honor codes and policies where students are expected to turn in fellow students for cheating also appear to have an important impact on cheating.

NOTES

1. For additional references to the crime deterrence literature, to which this article belongs, see Gneezy and Rustichini (2000: 2, n. 4).
2. Graves and Sexton have used this statute/policing trade-off model in other research (Graves, Lee, and Sexton 1989, 1990). It is closely related to the seminal Becker (1968) contribution.
3. We let T be sufficiently low so that if the cheating level of T were perfectly enforced, the result would be an average cheating level of C, $C < T$, where $B'(C) - E'(C) > 0$.

REFERENCES

Becker, G. S. (1968). "Crime and Punishment: An Economic Approach." *Journal of Political Economy* 2:76, 169–217.

Bunn, D. N., S. B. Caudill, and D. M. Gropper. (1992). "Crime in the Classroom: An Economic Analysis of Undergraduate Student Cheating Behavior." *Journal of Economic Education* 23, Summer: 197–207.

Collison, M. (1990). "Survey at Rutgers University Suggests that Cheating May Be on the Rise at Large Universities." *Chronicle of Higher Education* 37:A31–32.

Graves, P. E., D. R. Lee, and R. L. Sexton. (1989). "Statutes versus Enforcement Costs: The Case of the Optimal Speed Limit." *American Economic Review* 79:4.

———. (1990). "A Note on Drinking Driving and Enforcement Costs." *Southern Economic Journal* 56:3.

Gneezy, U., and A. Rustichini. (2000). "A Fine Is a Price." *Journal of Legal Studies XXIX*:1–17.

3

Culture

School Culture: Exploring the Hidden Curriculum

David J. Wren

Educators frequently overlook school culture. This article encourages teachers and administrators to gain a more complete picture of the school environment through an exploration of the symbolic nature of the hidden, or implicit, curriculum. A historical overview of the influence of the hidden curriculum on the educational process is presented. In addition, a checklist for examining symbolic aspects of the school environment is provided.

Since the mid-1970s (the era of human resources management), the study of behavior in organizations has strongly influenced the practice of school administration. Previously, administrative theory had stressed a "scientific" approach to educational goals, setting forth objectives in explicit, behavioral terms. However, planning systems, including management by objectives (MBOs) and planning, programming, and budgeting systems (PPBSs), offshoots of the scientific approach in the management of educational institutions and governmental agencies, often underestimated the importance of the dynamics of human interactions in organizational behavior.

In order to have a more complete picture of their schools, administrators need to become cognizant of the almost imperceptible yet powerful influence of institutional culture/climate. Culture refers to the values and symbols that affect organizational climate (in this case, students' and educators' perceptions of the school environment). According to Owens (1987), the symbolic aspects of school activities (e.g., traditions, rites, and rituals) must be considered, for these are "the values that are transmitted literally from one generation of the organization to another" (p. 168). The present article explores the hidden, or implicit, curriculum-school spirit, or ethos-as well as its beneficial and detrimental effects on the teaching/learning environment.

THE HIDDEN CURRICULUM UNCOVERED

Usually, when educators refer to school curriculum, they have explicit, consciously planned course objectives in mind. In contrast to this didactic curriculum, students experience an "unwritten curriculum" characterized by informality and lack of conscious planning. In fact, "all students must internalize a specific program of social norms for training in order to function effectively as members of a smaller society, the school, and later on as productive citizens of the larger American society" (Wren, 1993, p. 3). Thus, teachers' and administrators' interactions with students help shape attitudes and ideals (Henry, 1955).

Historical Overview

The two curricula were united in American classrooms from colonial times until the late 19th century. The school environment was carefully supervised by teachers and administrators, who expected conformity both in behavior and academics. Hirsch (1987) found almost complete congruence of values within American schools during this era. Ryan (1987) has described how the McGuffey graded reader series was used to inculcate discipline, good conduct, punctuality, respect for authority, and other commonly held social values.

During the post–Civil War period, instruction consisted mostly of transmitting factual information to rows of quiet, submissive students, many of whom were recent immigrants. Thus, America's public schools functioned much like a factory (Apple & King, 1983).

From the late-19th to mid-20th century, progressive educators, such as John Dewey, William Kilpatrick, and Harold Rugg, helped to bring about major changes. Religious teachings, so common in the previous century, were largely removed from the public schools (Ryan, 1987). Vallance (1973) concluded that, as a direct result of this progressivism, teachers became uncomfortable with their traditional role as inculcators of values. Instead, they relied on the school environment to be the socializing agent for overall student development.

The Hidden Curriculum's Impact on Behavior

During the past three decades, researchers have investigated both the beneficial and detrimental effects of school climate on the socialization process. Regarding positive effects, Bloom (1981) and Baltzell (1979) have described how the two curricula have worked in a complementary fashion (e.g., in Quaker schools). Kraybill (1991) has described how a school for Mennonites (a religious sect that traditionally encouraged separation from worldly affairs) passed on the faith as well as began a program of active involvement in community issues. Jackson (1968), a pioneer in the study of the hidden curriculum, reported that valuing successful competition in the working world had an effect on students' skills, beliefs, and attitudes toward work.

In terms of negative effects, discipline problems may occur for students who have difficulty following and internalizing classroom rules and daily routines (Jackson, 1968). The hidden curriculum can also promote student reluctance to challenge teachers on educational issues. In addition, Sadker and Sadker (1985) found that boys received more of their teachers' time and attention, whereas girls often were more docile and reticent to call attention to themselves.

Paradoxes

There are several paradoxes regarding the nature and utility of the hidden curriculum in schools. First, since it is by nature more spontaneous and less explicit than the regular curriculum, is there anything educators can do to modify its influence on students? Second, if it is possible for educators to achieve consistency in its application, will shedding its hidden nature alter its influence? Finally, is it desirable for educators to seek renewed uniformity in school culture in a pronouncedly multicultural society?

HIDDEN CURRICULUM CHECKLIST

Attaining greater understanding of the hidden curriculum offers a grassroots approach to complement top-down school improvement methods (e.g., MBOs and PPBSs). The following checklist can help educators examine the symbolic aspects of the school environment.

I. School Rules, Ceremonies, Rituals, and Routines

1. There are regular inter- and intrascholastic competitions, pep rallies, and schoolwide assemblies (yes or no).
2. There are opening convocations and appropriate end-of-the-year ceremonies and activities (yes or no).
3. The school has its own motto, colors, etc. (yes or no).
4. There are regularly scheduled field trips (yes or no).
5. Students regularly receive recognition for outstanding conduct, grades, and other achievements (yes or no).
6. School policies regarding homework, discipline, and safety (e.g., fire drills) are well-known to both faculty and students, and are consistently enforced by the principal (yes or no).

II. Document Analysis

1. Documents available for students' use (check each type): (a) yearbook, (b) school newspaper, (c) handbook, (d) school calendar.
2. Documents available for faculty and community members (check each type): (a) handbook, (b) announcements, (c) mission statement, (d) newsletters (e) reports on school/community service projects.

CONCLUSIONS

Educators need to be aware of the symbolic aspects of the school environment (i.e., its culture), as well as adolescents' and teachers' perceptions (i.e., school climate). Greater understanding of the hidden curriculum will help them to achieve the goal of providing effective schools in the 21st century.

REFERENCES

Apple, M., & King, J. (1983). *Humanistic education.* Berkeley, CA: McCutchon.

Baltzell, E. D. (1979). *Puritan Boston and Quaker Philadelphia: Protestant ethics and the spirit of class authority and leadership.* New York: Macmillan.

Bloom, B. (1981). *All our children learning.* New York: McGraw Hill.

Henry, J. (1955). Docility or giving the teacher what she wants. *Journal of Social Issues, 11,* 41–53.

Hirsch, E. D. (1987). *Cultural literacy.* Boston: Houghton Muffin.

Jackson, P. (1968). *Life in classrooms.* New York: Holt, Rinehart, and Winston.

Kraybill, D. (1991). *Passing on the faith: The study of a Mennonite school.* Intercourse, PA: Good Books.

Owens, R. (1987). *Organizational behavior in education* (3rd ed.). New Jersey: Prentice-Hall.

Ryan, K. (1987). *Character development in the schools and beyond.* New York: Praeger.

Sadker, D., & Sadker, M. (1985). Sexism in the classroom. *Journal of Vocational Education, 60*(7), 30–32.

Vallance, E. (1973). Hiding the hidden curriculum. *Curriculum Theory Network, 4,* 5–21.

Wren, D. (1993). *A comparison of the theories of adolescent moral development of Lawrence Kohlberg and Carol Gilligan: Alternative views of the hidden curriculum.* Doctoral dissertation, Lehigh University.

4

Socialization

Racial Socialization and Racial Identity: Can They Promote Resiliency for African American Adolescents?

David B. Miller

Although there is a rich body of research on resiliency, much of the literature fails to include minority youths or does not take into consideration their distinctive racial and environmental circumstances. Additionally, limited attention has been given to protective factors that are unique to nonmajority populations. This article posits that racial socialization and racial identity protect urban African American adolescents against some of the harmful effects of a discriminatory environment. These factors are hypothesized to influence academic achievement—an indicator of resiliency that has been used in many studies. A theoretical framework is provided that combines character development in a hostile environment, bicultural identity, and urban stress models. Implications for practice and future research are discussed.

While the concept of resiliency and factors that promote it have received considerable attention in the social science literature, far fewer studies have examined the development of resiliency among members of racial minorities. This paper addresses the need to expand the concept of resiliency to include protective factors unique to African American adolescents, specifically racial socialization and racial identity.

First, racial socialization and racial identity are presented as protective factors for urban African American adolescents. Peters (1985) and Stevenson (1994, 1995) have posited that racial socialization can act as a buffer against negative racial messages in the environment. Arroyo and Zigler (1995) have found that racial identity facilitates the development of competencies among African American adolescents. It is argued here that protective factors unique to nonmajority populations must be considered when assessing group strengths.

Second, a theoretical framework undergirding this argument is provided. This theoretical perspective takes into account the distinctive environmental conditions of African Americans. A better understanding of resiliency and associated factors is thereby achieved.

Educational achievement has long been considered as signifying resiliency among adolescents. However, limited attention has been given to the factors that promote educational achievement among urban adolescents (Barbarin, 1993; Bowman & Howard, 1985), even though the literature is replete with deficit-based discussions on the factors contributing to educational failure. The relationship of racial socialization and racial identity to the educational involvement and academic achievement of African American adolescents is thus discussed.

Finally, directions for future research and service delivery are presented.

RESILIENCY

Although environmental disadvantage and stress can lead to behavioral and psychological problems among children (Luthar & Zigler, 1991), there are those who overcome these difficulties to become well-adjusted adults (Garbarino, Dubrow, Kostelny, & Pardo, 1992; Luthar & Zigler, 1991; Safyer, 1994). This positive adaptation despite negative environmental circumstances is referred to as resiliency. Research into resiliency has focused on protective factors that enable an individual to adapt successfully to the environment, notwithstanding challenging or threatening circumstances (Garmezy, 1991; Masten, Best, & Garmezy, 1990). Whereas initial research centered on the absence of psychopathology among those experiencing negative life events, the current focus is toward understanding the process of resiliency (Smith & Prior, 1995).

Resiliency may include an array of abilities or attributes. Referred to as the "positive pole" (Rutter, 1987, p. 316), "unusually good adaptation" (Beardslee, 1989, p. 267), "positive psychological adjustment" (Smith & Prior, 1995, p. 173), success in meeting developmental tasks or social expectations (Luthar & Zigler, 1991), and the ability to "thrive, mature, and increase competence" (Gordon, 1995, p. 239), resiliency is indeed a broadly defined concept. The ability to "bounce back, recover, or form a successful adaptation in the face of obstacles and adversity" (Zunz, Turner, & Norman, 1993, p. 170) appears to encapsulate these various definitions best.

Unfortunately, it is often difficult to limit the degree of stress that some individuals experience, particularly those in economically and socially disadvantaged environments. Some researchers have examined the concept of invulnerability in regard to the effects of stressful situations (Garmezy, 1991; Rutter, 1993). However, everyone experiences stress, which requires a degree of adjustment. Rutter (1993) has suggested that susceptibility to stress is a graded phenomenon (p. 626) in that some are able to handle or recover from stressful events more readily than others. As with immunizations, individuals may develop resiliency after exposure to negative events, which becomes evident in the presence of "obstacles, adversity, stress, and high risk" (Zunz et al., 1993, p. 171).

Researchers have posited that protective factors operate at three levels: individual, familial, and societal (Garmezy, 1985; Gordon, 1995; Luthar & Zigler, 1991; Rutter, 1987). These elements may interact to protect a person from negative environmental conditions (Brooks, 1994; Rutter, 1987). To a significant degree, the lack of these elements makes one vulnerable to negative outcomes. For example, an adolescent without sufficient parental monitoring or nurturing may be susceptible to the environmental forces that contribute to delinquent or self-destructive behavior.

Although research into resiliency and the factors associated with it has increased (Brooks, 1994; Smith & Prior, 1995), more studies are needed on racial minorities. Luthar, Doernberger, and Zigler (1993) have indicated that while urban youth are at-risk for multiple behavioral problems, few empirical investigations have been undertaken regarding resiliency within this group. Rutter (cited in Garmezy, 1985) indicates that "many children do not succumb to deprivation, and it is important that we determine why this is so and what is it that protects them from the hazards they face" (p. 217). How African American children are able to survive and thrive in the face of adversity clearly requires more attention (Barbarin, 1993).

The exploration of additional protective factors within populations that have unique stressors and histories is paramount for further understanding resiliency in general, and minority groups in particular. Specifically, researchers have identified racial socialization and racial identity as capable of protecting African Americans from the effects of a hostile environment, but few systematic studies have been undertaken to investigate how these factors operate (McCreary, Slavin, & Berry, 1996).

Racial Socialization and Racial Identity

African American children in urban settings often have numerous obstacles to overcome, such as poverty, substandard housing, and inferior schools (Peters, 1985; Safyer, 1994). In addition, socialization of African Americans frequently occurs in the context of racial discrimination and oppression (McCreary et al., 1996), an environment that is not conducive to mental health (Thornton, Chatters, Taylor, & Allen, 1990).

Peters (1985) defined racial socialization as the "tasks Black parents share with all parents—providing for and raising children . . . but [they] include the responsibility of raising physically and emotionally healthy children who are Black in a society in which being Black has negative connotations" (p. 161). Thornton et al. (1990) described racial socialization in terms of personal and group identity, intergroup and interindividual relationships, and position in the social hierarchy. It must be pointed out that not all African American parents socialize their children regarding racial issues and prejudice (Bowman & Howard, 1985; Stevenson, 1994), but this usually leaves these children vulnerable (Spencer & Markstrom-Adams, 1990). Thus, racial socialization can act as a buffer against a hostile environment (Stevenson, 1994).

The socialization process is not the same in all African American families. It can be direct or indirect, verbal or nonverbal, overt or covert (Stevenson, 1994; Thornton et al., 1990). It can transpire through the observation of "modes, sequences, and styles of behavior" (Boykin & Toms, 1985, p. 42) during interaction with family members. Modeling of behaviors and exposure to culturally relevant material and activities are some of the methods that parents can use to facilitate this process. Nonetheless, the critical message is that race will affect available options and chances of succeeding in life, and competencies to navigate a sometimes hostile environment must be developed. Specifically the acquisition of a good education was identified by parents in Peters' (1985) study as essential for success in mainstream society.

The literature has pointed to the family as essential to the development of resiliency (Garmezy, 1985, 1991; Rutter, 1987). Families transmit the values, norms, and beliefs that are needed by successive generations to cope in an environment in which race plays a critical role (Demo & Hughes, 1990). In a study of 377 African American youths, Bowman and Howard (1985) found that resiliency was promoted among academically achieving adolescents as a result of proactive socialization by their parents. These parents conveyed to their children the importance of ethnic pride and self-development, and an awareness of racial barriers.

Racial socialization in turn fosters racial identity. Helms (1990) defined racial identity as "one's perception that he or she shares a common racial heritage with a particular group" (p. 3). Cross, Parkham, and Helms (1991) have posited that one of the functions of racial identity is "to defend and protect a person from psychological insults, and, where possible, to warn of impending psychological attacks that stem from having to live in a racist society" (p. 328). Arroyo and Zigler (1995) have indicated that racial identity operates in a "multifaceted manner" (p. 904) to affect a person's behavior and psychological states.

Identity development is a major task for all adolescents. However, for adolescents who are members of racial or ethnic minorities, this task is particularly complicated given their environment (Spencer & Markstrom-Adams, 1990). Parkham (cited in McCreary et al., 1996) has noted that African American adolescents must develop a strong racial identity in order to overcome the stigma of negative social stereotypes. In a study of 297 African

American adolescents, McCreary et al. (1996) found that high racial identity was a significant factor in the successful handling of stress, as well as in the lower rate of participation in problem behaviors.

According to Stevenson (1995), racial identity develops through racial awareness. Racial awareness is facilitated by racial socialization (Plummer, 1995). In fact, Sanders Thompson (1994) found that racial identity, among 225 African-Americans, was significantly influenced by racial socialization. While racial socialization is an important factor in the development of racial identity, socialization is influenced by the racial identity of the family. Thus, racial socialization and racial identity are inextricably bound.

A few studies have shown that racial socialization and racial identity can buffer African American adolescents against negative or stressful environmental conditions (Bowman & Howard, 1985; Peters, 1985; Stevenson, 1994). Further research is required into the contribution of these variables (e.g., manner, degree) to the resiliency of African American adolescents.

Theoretical Perspective

Spencer and Markstrom-Adams (1990) have recommended a multifaceted approach to understanding the developmental processes of minority youth. In order to integrate racial socialization and racial identity into a resiliency perspective, a theoretical framework combining character development in a hostile environment, bicultural identity, and urban stress models is offered here.

Character development. Chestang (1972) has stated that the covert and overt racism experienced by African Americans affects their character development. Development is thus predicated upon three interdependent conditions: social injustice, societal inconsistency, and personal impotence. The hardships caused by social injustice lead to frustration over the discrepancy between American ideals and reality, as well as feelings of impotence at being unable to influence one's environment.

The developmental process can lead to one of two outcomes—depreciated character or transcendent character. With depreciated character, a sense of worthlessness, inadequacy, and impotence is incorporated into the extrinsically imposed devaluation of self (Chestang, 1972). The individual will likely turn away from or against societal institutions. Conversely, with transcendent character there is more optimism. The individual seeks to overcome environmental adversity. Nevertheless, this individual may experience alienation from other group members, fostering depreciated character. Chestang cautions that while behaviors associated with one character predominate, in some interpersonal domains the individual can exhibit behaviors linked to the other character.

Transcendent character may be manifested in the pursuit of academic achievement and the development of a strong sense of self, as well as the establishment of beneficial familial and community connections. These have been tied to racial socialization and racial identity.

Bicultural identity. Clark (1991) has noted that some African American adolescents adapt to a discriminatory environment by developing a bicultural identity. Specifically, this helps them to achieve academically while maintaining a strong sense of group membership. Gordon (1995) has pointed out that the cultural differences experienced by African Americans within the academic environment are great, and that a bicultural identity may assist them in overcoming this obstacle to success.

Urban stress. Myers' (1982) model incorporates race and social class into the analysis of the impact of stress on minority populations. This urban stress model has six basic components: (1) exogenic (i.e., external) and endogenic (i.e., internal) antecedents, (2) internal and external mediators, (3) eliciting stressor(s), (4) the stress state, (5) coping and adaptation process, and (6) health outcome (p. 123). Internal factors include a clear sense of self and group identification. The hostile or negative environment makes up the exogenic antecedent. Racial identity is an internal mediating factor, whereas racial socialization is an external mediating factor. Myers posits that these internal and external factors are important in the development of a stress-resistant (i.e., resilient) lifestyle.

The components of the urban stress model provide a framework through which resiliency of disadvantaged urban African American youths may be examined. According to this model, vulnerability is intimately connected to a social environment that perpetuates discrimination (e.g., in the educational and legal spheres). The stress caused by a hostile environment clearly affects the individual's degree of risk for disorders, but racial socialization and racial identity improve the disadvantaged African American adolescent's ability to cope. This, in turn, influences educational involvement and academic achievement.

DISCUSSION

Research on resiliency has often neglected a population for which overcoming challenging and adverse conditions is a constant activity, namely African Americans. Given that the concept of resiliency is premised upon the influence of protective factors, the inclusion of resources unique to this group is imperative. Through their effects, racial socialization and racial identity can enable African American adolescents to overcome the covert and overt obstacles present in a hostile environment.

While some African American families provide strong racial socialization and promote racial identity, others de-emphasize these factors. Research comparing adolescents from these two types of families would be enlightening. The manner in which African American families provide racial socialization and how these messages contribute to the development of racial identity also need to be studied. Additionally, the effects of sociodemographic factors and access to resources, both economic and social, on the processes of racial socialization and racial identity development require empirical investigation.

It is essential that service providers pay special attention to those factors that facilitate positive outcomes against a backdrop of racial discrimination and inequality. Thus, levels of racial socialization and racial identity are critical considerations for those providing assistance to African American families and adolescents (Stevenson, 1994).

It is recommended that service providers inculcate proactive strategies for maintaining a sense of self, explore how African American adolescents can adjust to life in two worlds (one black, one white), and design intervention and prevention programs that focus on cultural strengths. Prior to service delivery, psychosocial assessment should, of course, consider the developmental impact of living in a hostile environment.

It is critical that social researchers rigorously explore how racial socialization and racial identity promote resiliency among African American adolescents. What dimensions of each construct exert the most influence on the development of resiliency? In what areas do these factors promote resiliency (i.e., positive mental health, anger management)? The theoretical perspective presented here needs to be investigated in regard to interactions involving stress and academic achievement. Researchers also must focus on the development of measures that can tap into the various dimensions of resiliency and quantify them.

African American adolescents in general, and those in an urban environment in particular, are faced with myriad adverse messages. In the face of seemingly overwhelming obstacles, many nevertheless become well-adjusted, contributing members of society. Through the expansion of a strengths perspective (i.e., resiliency) to include unique protective factors, the emphasis on pathology and deficits can be countered.

REFERENCES

Arroyo, C. G., & Zigler, E. (1995). Racial identity, academic achievement, and the psychological well-being of economically disadvantaged adolescents. *Journal of Personality and Social Psychology, 69*(5), 903–914.

Barbarin, O. A. (1993). Coping and resilience: Exploring the inner lives of African American children. *Journal of Black Psychology, 19*(4), 478–492.

Beardslee, W. R. (1989). The role of self-understanding in resilient individuals: The development of a perspective. *American Journal of Orthopsychiatry, 59*(2), 266–278.

Bowman, P. I., & Howard, C. (1985). Race-related socialization, motivation, and academic achievement: A study of black youths in three-generation families. *Journal of the American Academy of Child Psychiatry, 24*(2), 134–141.

Boykin, A. W., & Toms, F. D. (1985). Black child socialization. In H. P. McAdoo & J. L. McAdoo (Eds.), *Black children* (pp. 33–54). Beverly Hills, CA: Sage Publications.

Brooks, R. B. (1994). Children at risk: Fostering resilience and hope. *American Journal of Orthopsychiatry, 64*(4), 545–553.

Chestang, L. (1972). *Character development in a hostile environment* (Occasional Paper No. 3). Chicago: University of Chicago, School of Social Service Administration.

Clark, M. L. (1991). Social identity, peer relations, and academic competence of African-American adolescents. *Education and Urban Society, 24*(1), 41–52.

Cross, W. E., Parkham, T. A., & Helms, J. E. (1991). The stages of black identity development: Nigrescence models. In R. L. Jones (Ed.), *Black psychology* (3rd ed., pp. 319–338). Berkeley, CA: Cobb & Henry.

Demo, D. H., & Hughes, H. (1990). Socialization and racial identity among Black Americans. *Social Psychology Quarterly, 53*(4), 364–374.

Garbarino, J., Dubrow, N., Kostelny, K., & Pardo, C. (1992). *Children in danger: Coping with the consequences of community violence.* San Francisco: Jossey-Bass.

Garmezy, N. (1985). Stress-resistant children: The search for protective factors. In J. E. Stevenson (Ed.), *Recent research in developmental psychopathology* (pp. 213–233). Oxford: Pergamon Press.

Garmezy, N. (1991). Resilience in children's adaptation to negative life events and stressed environments. *Pediatric Annals, 20*(9), 459–466.

Gordon, K. A. (1995). Self-concept and motivational patterns of resilient African American high school students. *Journal of Black Psychology, 21*(3), 239–255.

Helms, J. E. (1990). *Black and white racial identity: Theory, research and practice.* New York: Greenwood Press.

Luthar, S. S., Doernberger, C. H., & Zigler, E. (1993). Resilience is not a unidimensional construct: Insights from a prospective study of inner-city adolescents. *Development and Psychopathology, 5,* 703–717.

Luthar, S. S., & Zigler, E. (1991). Vulnerability and competence: A review of research on resilience in childhood. *American Journal of Orthopsychiatry, 61*(1), 6–21.

Masten, A. S., Best, K. M., & Garmezy, N. (1990). Resilience and development: Contributions from the study of children who overcome adversity. *Development and Psychopathology, 2,* 425–444.

McCreary, M. L., Slavin, L. A., & Berry, E. J. (1996). Predicting problem behavior and self-esteem among African American adolescents. *Journal of Adolescent Research, 11*(2), 216–234.

Myers, H. F. (1982). Stress, ethnicity, and social class: A model for research with black populations. In E. E. Jones & S. J. Korchin (Eds.), *Minority mental health* (pp. 118–148). Westport, CT: Praeger Publishers.

Peters, M. F. (1985). Racial socialization of young Black children. In H. P. McAdoo & J. L. McAdoo (Eds.), *Black children: Social, educational, and parental environments* (pp. 159–173). Newbury Park, CA: Sage.

Plummer, D. L. (1995). Patterns of racial identity development of African American adolescent males and females. *Journal of Black Psychology, 21*(2), 168–180.

Rutter, M. (1987). Psychosocial resilience and protective mechanisms. *American Journal of Orthopsychiatry, 57*(3), 316–331.

Rutter, M. (1993). Resilience: Some conceptual considerations. *Journal of Adolescent Health, 14,* 626–631.

Safyer, A. W. (1994). The impact of inner-city life on adolescent development: Implications for social work. *Smith College Studies in Social Work, 64*(2), 153–167.

Sanders Thompson, V. L. (1994). Socialization to race and its relationship to racial identification among African Americans. *Journal of Black Psychology, 20*(2), 175–188.

Smith, J., & Prior, M. (1995). Temperament and stress resilience in school-age children: A within-families study. *Journal of the American Academy of Child and Adolescent Psychiatry, 34*(2), 168–179.

Spencer, M. B., Cole, S. P., DuPree, D., Glymph, A., & Pierre, P. (1993). Self-efficacy among urban African American early adolescents: Exploring issues of risk, vulnerability, and resilience. *Development and Psychopathology, 5,* 719–739.

Spencer, M. B., & Markstrom-Adams, C. (1990). Identity processes among racial and ethnic minority children in America. *Child Development, 61,* 290–310.

Stevenson, H. C. (1994). Racial socialization in African American families: The art of balancing intolerance and survival. *The Family Journal: Counseling and Therapy for Couples and Families, 2*(3), 190–198.

Stevenson, H. C. (1995). Relationship of adolescent perceptions of racial socialization to racial identity. *Journal of Black Psychology, 21*(1), 49–70.

Thornton, M. C., Chatters, L. M., Taylor, R. J., & Allen, W. R. (1990). Sociodemographic and environmental correlates of racial socialization by black parents. *Child Development, 61,* 401–409.

Zunz, S. J., Turner, S., & Norman, E. (1993). Accentuating the positive: Stressing resiliency in school-based substance abuse prevention programs. *Social Work in Education, 15*(3), 169–176.

5

Society, Social Structure, and Interaction

Longitudinal Perspective: Adverse Childhood Events, Substance Use, and Labor Force Participation Among Homeless Adults

Tammy W. Tam, Cheryl Zlotnick, and Marjorie J. Robertson

Objectives: We examined the long-term effects of adverse childhood events on adulthood substance use, social service utilization, and subsequent labor force participation. Methods: A county-wide probability sample of 397 homeless adults was interviewed three times in a 15-month period. By using a path model, literature-based relationships between adverse childhood events and labor force participation with the mediating effects of adulthood substance use and service use were tested.

Results: Adverse childhood events were precursors to adulthood alcohol and drug use. Consistent substance use was negatively associated with long-term labor force participation and with social service utilization among homeless adults. Adverse events at childhood, however, were positively associated with service use.

Conclusions: Adverse childhood events may contribute to negative adulthood consequences, including consistent substance use and reduced labor force participation. Agencies that are involved in halting the abuse or neglect also should participate in more preventive interventions. Job-related assistance is particularly important to facilitate employment and labor force participation among homeless adults.

INTRODUCTION

Work and Homeless Adults

In western society, work provides income, contributes to one's self-identity, delineates social status, adds structure to the day, and provides opportunities for social connections (1, 2). Conversely, unemployment may not only indicate the loss of adulthood roles and decreased income but also a lack of connection with mainstream society. Homeless adults often are perceived as living outside mainstream society; consequently, many programs for homeless adults use work-related outcome indicators to demonstrate successful treatment and exits from homelessness (3–5).

Yet, a recent study reported that as many as 87% of homeless adults are either working or seeking work (i.e., participating in the labor force) (6), although the jobs are sporadic, rarely include benefits, and pay poorly (7–10). More importantly, although the majority may be working or seeking work at any point in time, fewer than half report working consistently for periods of at least 12 consecutive months (6).

Although there are few definitive studies on the work lives of homeless adults, many demonstration projects that target homeless adults offer explanations for their episodic work histories (3). First, many homeless adults are unskilled; thus their employability is limited, particularly if the unemployment rate is up, and the national economy is troubled. Work experience and knowledge about appropriate work behaviors are limited. But probably the most noted explanations concern the prevalence and severity of mental illness and substance abuse.

Substance Abuse and Mental Illness

Compared with the general population, a disproportionate number of homeless adults have substance abuse problems (11–14) and/or mental health disorders (12, 15, 16). Even in the general (nonhomeless) adult population, those with severe mental illness are less likely to be employed compared with those without severe mental illness (17).

As with mental illness, adults in the general population with substance abuse problems are less likely to be employed than those without substance abuse problems (18, 19). A similar finding was noted with homeless adults. Those who reported recent drug use compared with those without were significantly less likely to be in the labor force (6).

In the past few decades, a focused attempt to help homeless adults rejoin the workforce has been made through a series of federally funded demonstration projects (3, 20). Although researchers anticipated the need for mental health and substance abuse treatment, they were unprepared for other difficulties. For example, many demonstration projects noted that in addition to few marketable skills, low education, untreated mental illness and untreated substance use, job readiness (i.e., an understanding of workplace expectations) pose the greatest barrier to job attainment and maintenance (3). However, the motivation and interest in excelling at the workplace are no different for homeless adults than for other unemployed and employed adults (21).

Adverse Childhood Events

No longitudinal studies are available to examine the link between childhood characteristics and labor force participation. However, there is abundant research noting the difficult circumstances that many homeless adults experienced as children, including sexual and physical abuse (22, 23) and out-of-home placements, including foster care (22, 24–28). Among homeless adults, the prevalence rates of childhood sexual and physical abuse are high relative to the general population, and rates vary from 13%–27% in samples of homeless adults drawn from New York and California (29–31).

Several studies have noted an association between reported childhood sexual and physical abuse and adulthood substance abuse in men and women and in homeless and nonhomeless samples (23, 31–34). For example, almost 25% of homeless women with substance abuse problems reported childhood sexual and physical abuse (23).

While the associations among adverse childhood events, substance abuse, mental illness, and unemployment are clear, little research examines the chain of events leading to labor force participation, and none includes adverse childhood events, substance abuse, and mental illness. Another important factor is service-agency interventions that target homeless adults. Other than the demonstration projects mentioned earlier, there is little information on the impact that services may have on entry into the labor force. Therefore, the two research questions of this study are:

1. Among homeless adults, are adverse childhood events precursors to a chain of adulthood conditions including heavy substance abuse and the lack of labor force participation?
2. Does use of social services contribute to entry into the labor force?

METHODS

Sample

This study used an existing longitudinal data set called the Study of Alameda County Homeless Residents (STAR) Project. The methodology used to sample and collect data for the STAR Project is summarized here and described elsewhere in more detail (13, 35–37).

The target population included homeless adults who were at least 18 years old and who spent the previous night in one of the following situations: 1) on the streets (i.e., in unconventional accommodations, including abandoned or public buildings, vehicles, or out of doors); 2) in an emergency shelter or mission; 3) in a hotel or motel where rent was paid with a voucher from a social service agency; or 4) doubled up with friends or family if the individual also had been literally homeless at least one night of the past 30 days (i.e., spent the night on the streets, in a shelter, or in a vouchered hotel room).

To collect a representative sample of homeless adults, all sites in Alameda County that regularly provided free meals at least once per week were included in the sampling frame. Eighty sites were identified, consisting of emergency shelters (both individual and family), meal programs, and drop-in centers. The sampling frame excluded jails, hospitals, residential treatment facilities, and domestic violence shelters (36).

The sampling design was a multistage cluster sample with stratification. Clusters were based on the type of site (i.e., meal program or shelter), and selection at each stage was stratified based on size and location (i.e., inside vs. outside the city of Oakland). The three stages of the multistage cluster design included: 1) selection of 29 of 80 sampling sites including shelters, meal programs, and drop-in centers with a probability proportionate to size (PPS); 2) selection of the day and mealtime for interviews at each site (also PPS); and 3) random selection of respondents at each site. Weights were calculated to adjust for the varying probabilities of each respondent being selected into the sample (36).

Interviewers received about 40 hours of classroom training on the administration of the standardized instruments and sampling criteria. Respondents were paid $25 to complete the baseline interview that averaged two hours and $20 for each of the two follow-up interviews.

The sampling strategy was executed within a 30-day period and repeated three times between April and August 1991. A total of 564 adults were recruited into the baseline sample (Wave 1), with an overall completion rate of 90.4%. No differences in gender or race/ethnicity were found between those who agreed to be interviewed and those who refused. Five-month follow-up interviews were completed for 473 (83.9%) of the baseline sample. Fifteen month follow-up interviews (Wave 3) were completed for 397 (70.4%) of the baseline sample. There were no significant differences between the baseline and longitudinal samples for the following key demographic characteristics: gender, household composition, age, education, race/ethnicity, duration of homelessness or diagnostic status (37). The longitudinal sample consisted of 397 individuals; however, we excluded 20 individuals who had missing data on employment for the third wave interview. One person who qualified for the diagnosis of alcohol or drug use disorder solely based on the job criteria also was excluded to avoid dependence between the independent and dependent variables. The final longitudinal sample consisted of 376 subjects.

Variables

Demographic data were collected at baseline. Age, education, race/ethnicity, and marital status were self-reported, while gender was attributed by interviewers. Chronic homelessness was measured at baseline by asking respondents to estimate the total time spent homeless since age 18.

Childhood events consisted of several dichotomous questions asked about respondent childhoods at baseline. Adverse childhood events referred to the respondents' experiences when they were under age 18, including a history of living in a foster care, group home or other out-of-home placement (such as living with friends, family members or an institution), sexual abuse, physical abuse, running away from home for seven nights or more, being arrested, homelessness, and early regular alcohol or drug use. The frequency of each adverse childhood event was examined separately in bivariate analyses. A new variable was created to indicate the number of adverse childhood events that were reported by each respondent. The aggregated variable "adverse childhood events" was used in the path model.

Substance use disorders were assessed at baseline by using the Diagnostic Interview Schedule, Version III Revised (DIS-III-R). The DIS-III-R provides a standardized structured interview designed for use by nonclinicians and is based on diagnostic criteria from the Diagnostic and Statistical Manual of the American Psychiatric Association, Version III-R (38). Diagnostic assessment included major mental disorders (i.e., major depression, bipolar disorder, and schizophrenia) and substance use disorders (i.e., alcohol and nine other categories of drugs). Respondents were classified as having a current disorder if they met diagnostic criteria for a lifetime diagnosis of an alcohol or drug use disorder and also had at least one symptom of the same disorder during the previous 12 months (39).

Substance Abuse Severity was measured in two ways. First, at each wave, respondents were asked if they drank five or more drinks of alcohol in a day in the past 30 days. Similarly, respondents were asked if they used drugs in the past 30 days. A dichotomous variable was created to indicate substance use (5 + alcohol or any drug use) in at least one interview. This variable was used in the bivariate comparisons. In addition, an ordinal variable "consistent substance abuse" was created to indicate whether respondents met either the alcohol severity criteria or the drug severity criteria at all three interviews, only at two interviews, only at one interview, or never.

Social Services used by respondents were measured by the following three items in each interview: 1) the number of agencies where the respondent was being assisted by a counselor, advocate, or case manager; 2) the number of types of assistance the agencies or advocates provided; and 3) whether respondents received help applying for entitlements, including Aid to Families with Dependent Children, Supplemental Security Income or Social Security Disability Income, and General Assistance. Functions served by agencies were grouped into three major categories: 1) help with housing; 2) help with job training and job search; and 3) help with other areas, including financial assistance/welfare, providing treatment, child care, food/clothing, legal assistance, and transportation. Any services received in each category were counted as one function, and the number of functions was summed.

Labor Force Participation. As with some recent studies (17, 18), the United States Department of Labor (DOL) definition was used in which labor force participation consists of all individuals who were currently employed or unemployed and had been seeking work for the past four weeks. Those who were unemployed, retired, or not seeking work (such as disabled or retired persons) were categorized as not participating in the labor force (40).

ANALYSIS

All analyses were conducted by using SPSS(R) (41) and Mplus (42). The chi-square test of independence was used to compare frequencies. The Fishers exact test (43) was used to compare contingency tables with small cell sizes. Multiple comparisons were adjusted by the Bonferroni method. The proposed path model was estimated by weighted least squares with robust standard errors and the mean- and variance-adjusted chi-square test statistic using Mplus. Model fit was evaluated by using summary statistics, including model chi-square, root mean square of approximation (RMSEA), and weighted root mean square residual (WRMR). As a general guideline, a model is considered to be a good representation of the data when the chi-square to degree-of-freedom ratio is less than two (44), RMSEA values are less than 0.05 (45, 46), or the WRMR value is less than 0.9 (46). The significance, direction, and magnitude of the regression paths in the model also were examined to evaluate how well the data fit the conceptual model. Significance of coefficients was evaluated as z statistics at $p = 0.05$. To obtain a parsimonious model, insignificant paths were fixed to zero for reestimation. The final model included only significant regression coefficients.

RESULTS

Sample Characteristics

Of the 376 homeless persons who were included in the final sample, the majority were male (74.7%), over 35 years old (61.0%), African American (66.9%), and married at least once (52.3%). A vast majority had lifetime alcohol or drug use disorder (70.7%), reported substance use in at least one wave (89.2%), and had experienced one or more negative childhood events (72.5%). About 60% of the sample had received services from agencies, and a similar percentage was in the labor force at the end of the 15-month follow-up period.

Bivariate Analyses

The mediators (heavy substance use and number of services used) and the outcome variable (labor force participation at the third wave interview) were compared with various background characteristics (see Table 1). While

Table 1. Sample characteristics of homeless adults by substance use, service utilization, and labor force participation (n = 376).

	Substance use (A)	
	Yes N (%)	No N (%N)
Total sample	324 (89.2)	52 (10.8)
Demographics and background		
Women	98 (22.5)	32 (48.8)***
Age 36+	188 (59.9)	33 (69.0)
African American	215 (69.0)	25 (48.8)*
Never married	156 (49.1)	19 (35.7)
Chronic homelessness	157 (54.8)	13 (35.0)*
Perceived physical or psychological disability	132 (38.0)	18 (42.5)
Substance use		
Lifetime alcohol or drug use disorder	244 (77.8)	5 (11.9)***
Lifetime major mental disorder	79 (25.7)	9 (14.6)
Substance use at one or more waves	—	—
Childhood events before age 18		
Had any adverse childhood event	248 (75.7)	28 (46.3)***
Ever lived in an out-of-home placement	137 (38.9)	18 (26.2)
Ever lived in foster care or group home	41 (9.4)	6 (10.9)
Childhood history of sexual abuse	53 (13.4)	7 (14.3)
Childhood history of physical abuse	42 (11.4)	5 (11.9)
Ran away from home seven nights or more	102 (29.2)	11 (23.8)
Arrested	132 (39.7)	6 (7.3)***
Homeless by self	55 (14.0)	5 (7.3)
Parents were on welfare	107 (33.9)	8 (21.4)
Social service use		
One or more types of help	199 (60.8)	31 (65.9)
Help with housing or shelter	132 (38.9)	22 (41.5)
Help with job training or job search	124 (40.6)	18 (35.7)
Other help	167 (50.6)	25 (52.4)
	Social service use (B)	
	Yes N (%)	No N (%N)
Total sample	230 (61.4)	146 (38.6)

(continued)

Table 1 (*Continued*)

	Yes N (%)	No N (%N)
Demographics and background		
Women	92 (30.5)	38 (16.9)**
Age 36+	134 (61.3)	87 (60.8)
African American	149 (68.9)	91 (63.8)
Never married	102 (42.6)	73 (56.1)**
Chronic homelessness	97 (47.6)	73 (60.8)*
Perceived physical or psychological disability	91 (40.2)	59 (35.8)
Substance use		
Lifetime alcohol or drug use disorder	166 (76.6)	83 (61.5)***
Lifetime major mental disorder	64 (32.2)	24 (12.8)***
Substance use at one or more waves	199 (88.5)	125 (90.5)
Childhood events before age 18		
Had any adverse childhood event	174 (75.4)	102 (7.8)
Ever lived in an out-of-home placement	98 (40.7)	57 (32.4)
Ever lived in foster care or group home	30 (10.2)	17 (7.4)
Childhood history of sexual abuse	42 (14.8)	18 (10.8)
Childhood history of	32 (14.0)	14 (7.4)
Ran away from home seven nights or more	71 (30.6)	42 (25.7)
Arrested	87 (39.6)	51 (30.4)
Homeless by self	35 (13.1)	25 (14.2)
Parents were on welfare	75 (32.8)	40 (32.4)
Social service use		
One or more types of help	—	—
Help with housing or shelter	—	—
Help with job training or job search	—	—
Other help	—	—

	Labor force status (c)	
	In N (%)	Out N (%)
Total sample	217 (60.3)	159 (39.7)
Demographics and background		
Women	69 (21.1)	61 (31.6)*
Age 36+	117 (54.5)	104 (71.1)***
African American	141 (69.3)	99 (63.2)
Never married	109 (51.9)	66 (41.4)
Chronic homelessness	92 (50.7)	78 (55.8)
Perceived physical or psychological disability	61 (25.1)	89 (58.9)***
Substance use		
Lifetime alcohol or drug use disorder	138 (69.7)	111 (72.4)
Lifetime major mental disorder	50 (24.2)	38 (25.5)
Substance use at one or more waves	185 (60.5)	139 (39.5)

Table 1 (*Continued*)

	IN N (%)	OUT N (%)
CHILDHOOD EVENTS BEFORE AGE 18		
Had any adverse childhood event	156 (73.2)	120 (71.7)
Ever lived in an out-of-home placement	86 (35.9)	69 (39.9)
Ever lived in foster care or group home	26 (7.4)	21 (12.4)
Childhood history of sexual abuse	30 (10.4)	30 (17.6)
Childhood history of physical abuse	28 (11.3)	19 (11.8)
Ran away from home seven nights or more	69 (32.0)	44 (23.7)
Arrested	73 (34.2)	65 (38.8)
Homeless by self	37 (14.3)	23 (12.5)
Parents were on welfare	58 (31.2)	57 (34.9)
SOCIAL SERVICE USE		
One or more types of help	129 (60.6)	101 (62.5)
Help with housing or shelter	81 (36.2)	73 (44.1)
Help with job training or job search	89 (45.0)	53 (32.7)*
Other help	102 (47.6)	90 (55.9)

Percentages are weighted; counts are unweighted.
(a) Reported at one or more waves: alcohol (5 or more drinks in a day) or drug use in the past 30 days.
(b) Reported at one ore more waves: use of one or more types of services.
(c) Labor force status at wave 3.
***$p \leq 0.001$.
**$p \leq 0.01$.
*$p \leq 0.05$.

women were less likely to report substance use in at least one interview ($p \leq 0.001$) and participation in the labor force ($p \leq 0.031$) at the 15-month follow-up, more women had received one or more types of social services ($p \leq 0.004$). African Americans ($p \leq 0.015$), those with chronic homelessness ($p \leq 0.028$), lifetime substance use disorder ($p \leq 0.000$), or one or more adverse childhood events ($p \leq 0.000$) [being arrested before age 18 in particular ($p \leq 0.000$)] reported higher rates of substance use. The number of services used was less for those who had experienced chronic homelessness ($p \leq 0.017$) or who had never been married ($p \leq 0.013$). In contrast, homeless adults who had lifetime substance use disorder ($p \leq 0.002$) or a major mental disorder ($p \leq 0.000$) reported using more types of services. Participation in labor force was higher among those who were younger ($p \leq 0.002$), had no perceived disability ($p \leq 0.000$), or had received help with job training or job search ($p \leq 0.021$).

Path Model

The path model was specified based on the following hypotheses: 1) adverse childhood events and consistent substance use will have direct negative long-term effects on labor force participation, whereas social service utilization will be positively related to labor force participation; 2) adverse childhood events will have direct positive effects on consistent substance use and service use; and 3) consistent substance use will be associated with lower levels of social service utilization (see Figure 1). (All figures referenced can be found within the online version of this article, at http://www.infotrac-college.com.) The model also controlled for the effects of gender, age, chronic homelessness, lifetime substance use and mental disorder, and perceived disability on the relationships detailed in the model. Consistent substance use was operationalized by an ordinal variable indicating alcohol (5 or more drinks in a day) or drug use in the past 30 days at none, one, two, or all three waves of interview. Similarly, service utilization was coded as an ordinal variable indicating the number of social service functions received over the 15-month period. All other variables in the model were coded as dichotomous variables.

The final model demonstrated good fit of the model to the data (χ = 10.52, p = 0.23, RMSEA = 0.03, WRMR = 0.69). Table 2 summarizes the estimated parameters for the regression of each outcome variable on the background variables included in the model.

Adverse childhood event had a positive direct effect on consistent alcohol or drug use (standardized β = 0.09) and social service utilization (β = 0.11). Although adverse childhood event had no direct effect on labor force participation, a significant negative path from consistent substance use to labor

Table 2. Standardized regression parameters and critical ratios (in parenthesis) for path model

	Substance Use (A)	Social Service Use (B)
Women	−0.1 (−2.38)	0.12 (2.57)
Age 36+	0.09 (2.28)	−0.02 (−0.48)
Chronic homelessness	0.17 (4.50)	0.05 (1.36)
Perceived physical or psychological disability	—	—
Lifetime alcohol or drug use disorder	0.43 (10.58)	0.34 (6.04)
Lifetime major mental disorder	—	0.17 (3.48)

	Labor Force Participation (C)
Women	−0.24 (−4.07)
Age 36+	−0.17 (−3.04)
Chronic homelessness	—
Perceived physical or psychological disability	−0.38 (−6.84)
Lifetime alcohol or drug use disorder	—
Lifetime major mental disorder	—

— Not included in final model.
(a) Reported at one or more waves: alcohol (5 or more drinks in a day) or drug use in the past 30 days.
(b) Reported at one or more waves: use of one or more types of services.
(c) Labor force status at wave 3.

force participation ($\beta = -0.14$) demonstrates the mediating effect of consistent substance use on this relationship. Consistent substance use also was inversely related to service utilization ($\beta = -0.29$). The effect of service use on labor force participation was not significant when controlling for the effects of other variables in the model.

With regard to the regression of the mediators and outcome variable on the background characteristics (Table 2), women were less likely to report consistent substance use ($\beta = -0.10$) or to participate in the labor force ($\beta = -0.24$). However, women reported getting help on more types of social services ($\beta = 0.12$). Those who were older than age 35 ($\beta = 0.09$), had been homeless for over a year ($\beta = 0.17$), or had lifetime alcohol or drug use disorder ($\beta = 0.43$) were more likely to use alcohol or drugs over the 15-month period than others. Services used over the three waves were higher for those who were younger ($\beta = -0.02$), had been homeless for over a year ($\beta = 0.05$), or had a lifetime substance use ($\beta = 0.34$) or major mental disorder ($\beta = 0.17$). Those who were age 35 or younger ($\beta = -0.17$) or had a perceived disability ($\beta = -0.38$) were less likely to be in the labor force.

CONCLUSIONS

This study examined whether selected adverse childhood events were related to adulthood behaviors among homeless adults. Results of our path model indicate that adverse childhood events are precursors to serious alcohol and drug use in adults, and that consistent substance use was negatively associated with long-term labor force participation among homeless adults.

Although this study contributes to the literature by creating and examining a model that links childhood events to adulthood characteristics among homeless adults, many studies have provided the underpinnings for this study. For example, studies of nonhomeless adults have noted the association between adverse childhood events and addiction to substance abuse (32, 47). In particular, studies indicate that there may be sequelae of childhood sexual and physical abuse in the form of mental illness or alcohol and drug addiction (23, 34, 47, 48).

This study also sought to assess opportunities for intervention to break the link between adverse childhood events and negative adulthood behaviors. The first opportunity is after a childhood event has occurred. Findings suggest that adverse childhood events were positively related to service use for homeless adults. Some childhood events (such as foster care, group home placement, or history of childhood arrests) involve contact with the juvenile justice and the child welfare systems. These experiences may facilitate other types of service contact and utilization among these same persons as adults. This finding also suggests the opportunity to intervene and ameliorate some of the negative consequences of these adverse childhood events that may lead to substance abuse or other negative life-altering behaviors. For example, the child welfare system typically responds to episodes of child neglect or abuse by removing children from the negative situation and placing them into foster care (49). This intervention may halt the episode of ongoing neglect or abuse, but it often provides no assessment of the psychological damage inflicted by the abuse and no preventive intervention to reduce or ameliorate later adulthood consequences that may emanate from unresolved psychological issues (23, 34, 48, 50). Thus, findings here support the increasing evidence that adverse childhood events may contribute to later adulthood consequences; and they suggest that agencies involved in halting the abuse or neglect of children also should participate in preventive interventions.

The path model created in this study also suggests that consistent substance abuse is negatively related to labor force participation. This finding is consistent with other studies that have examined this relationship in nonhomeless samples (18, 19). One explanation for this relationship is that substance use may impair work performance and increase absenteeism.

[GRAPHIC OMITTED] The path model also identified a negative association between consistent substance use and social service utilization. Avoidance of social services and treatment agencies by homeless adults with severe substance abuse problems has been noted elsewhere (51–53). The relationship often is attributed to a fear of prosecution for use of illicit drugs. Consequently, many agencies that target homeless adults have created two types of workers to address this problem. The first is an outreach worker, whose role is to identify individuals who need treatment but who may hesitate to seek treatment on their own (54, 55). The second is a case manager who coordinates a variety of services to meet their needs (56–58).

Among the services provided by case managers are linkages to employment. There is evidence to suggest these efforts have been successful. Our bivariate analyses indicated that among homeless adults, labor force participation was more likely among those who had obtained job-related assistance. Programs that target homeless adults with the aim of increasing labor force participation have acknowledged some of the difficulties of working with this population (3). Yet, findings here suggest that job-related assistance may facilitate workforce participation.

In conclusion, while findings do not show causal relationships, they do identify a series of associations that begin with adverse events at childhood, and they support studies that suggest the importance and influence of childhood events on adult behavior and experiences.

Readers are cautioned that these findings are based on self-reported retrospective data from homeless adults recruited from shelters and meal programs in a single county, and they are subject to recall bias. Findings may not generalize to homeless adults elsewhere.

ACKNOWLEDGMENTS

This work was supported by a grant from the National Institute on Alcohol Abuse and Alcoholism (AA12019). Data collection was supported by a grant from the National Institute of Mental Health (MH46104).

REFERENCES

1. Bahr HM. *Skid Row: An Introduction to Disaffiliation*. New York: Oxford University Press, 1973.
2. Smith R. "I feel really ashamed": how does unemployment lead to poorer mental health? *BMJ* 1985; 291:1409–1411.
3. Bailis LN, Blasinsky M, Chesnutt S, Tecco M. *Job Training for the Homeless: Report on Demonstration's First Year*. Rockville, MD: R.O.W. Sciences, 1991.
4. Lam JA, Rosenheck R. Correlates of improvement in quality of life among homeless persons with serious mental illness. *Psychiatr Serv* 2000: 51(1):116–118.
5. Rosenheck R, Morrissey JL, Calloway M, Stolar M, Johnsen M, Randolph F, Blasinsky M, Goldman HH. Service delivery and community: social capital, service systems integration, and outcomes among homeless persons with severe mental illness. *Health Serv Res* 2001; 36(4):691–710.
6. Zlotnick C, Robertson MJ, Tam T. Substance use and labor force participation among homeless adults. *Am J Drug Alcohol Abuse* 2002; 28(1):37–53.
7. Benda BB. Crime, drug abuse and mental illness: a comparison of homeless men and women. *J Soc Sci* 1990; 13(3):39–60.
8. Grigsby C, Baumann D, Gregorich SE, Roberts-Gray C. Disaffiliation to entrenchment: a model for understanding homelessness. *J Soc Issues* 1990; 46(4):141–156.
9. Koegel P, Burnam MA, Farr RK. Subsistence adaptation among homeless adults in the inner city of Los Angeles. *J Soc Issues* 1990; 46(4):83–107.
10. Zlotnick C, Robertson MJ. Sources of income among homeless adults with major mental disorders or substance use disorders. *Psychiatr Serv* 1996; 47(2):147–151.
11. Bray RM, Marsden ME. Prevalence of use of illicit drugs, alcohol, and cigarettes among DC metropolitan area household residents in 1990. The 100th Annual Meeting of the American Psychological Association, Washington, DC, August 14–18, 1992; 1992.

12. Lehman AF, Cordray DS. Prevalence of alcohol, drug and mental disorders among the homeless: one more time. *Contemp Drug Probl* 1993; 20(3):355–383.
13. Robertson MJ, Zlotnick C, Westerfelt A. Drug use disorders and treatment contact among homeless adults in Alameda County. *Am J Public Health* 1997; 87(2):221–228.
14. National Institute of Drug Abuse. Prevalence of Drug Use in the Washington, DC, Metropolitan Area Homeless and Transient Population: 1991. Washington, DC, Metropolitan Area Drug Study; Washington, DC: US Government Printing Office, 1993.
15. Fischer PJ. Estimating the prevalence of alcohol, drug and mental health problems in the contemporary homeless population: a review of the literature. *Contemp Drug Probl* 1989; 16(3):333–389.
16. Koegel P, Burnam MA, Farr RK. The prevalence of specific psychiatric disorders among homeless individuals in the inner-city of Los Angeles. *Arch Gen Psychiatry* 1988; 45(12):1085–1092.
17. Sturm R, Gresenz CR, Pacula RL, Wells KB. Labor force participation by persons with mental illness. *Psychiatr Serv* 1999; 50(11):1407–1408.
18. Bray JW, Zarkin GA, Dennis ML, French MT. Symptoms of dependence, multiple substance use, and labor market outcomes. *Am J Alcohol Abuse* 2000; 26(1):77–95.
19. Office of Applied Studies. Substance Use and Mental Health Characteristics by Employment Status. Rockville, MD: Substance Abuse and Mental Health Services Administration, 1999.
20. Bennett G, Shane P, Tutunijian BA, Perl HI. Job Training and Employment Services for Homeless Persons with Alcohol and Other Drug Problems. Department of Health and Human Services, 1992.
21. Porat H, Marshall G, Howell W. The career beliefs of homeless veterans: vocational attitudes as indicators of employability. *J Career Assess* 1997; 5(1):47–59.
22. Koegel P, Melamid E, Burman MA. Childhood risk factors for homelessness among homeless adults. *Am J Public Health* 1995; 85(12):1642–1649.
23. Nyamathi A, Longshore D, Keenan C, Lesser J, Leake BD. Childhood predictors of daily substance use among homeless women of different ethnicities. *Am Behav Sci* 2001; 45(1):35–50.
24. Bassuk EL, Buckner JC, Weinreb LF, Browne A, Bassuk SS, Dawson R, Perloff JN. Homelessness in female-headed families: childhood and adult risk and protective factors. *Am J Public Health* 1997; 87(2):241–248.
25. Herman DB, Susser ES, Struening EL, Link BL. Adverse childhood experiences: are they risk factors for adult homelessness. *Am J Public Health* 1997; 87(2):249–255.

26. Piliavin I, Sosin M, Westerfelt AH, Matsueda RL. The duration of homeless careers: an exploratory study. *Soc Serv Rev* 1993; 67:576–598.
27. Susser ES, Lin SP, Conover SA, Struening EL. Childhood antecedents of homelessness in psychiatric patients. *Am J Psychiatry* 1991; 148(8):1026–1030.
28. Zlotnick C, Robertson MJ, Wright MA. The impact of childhood foster care and other out-of-home placement on homeless women and their children. *Child Abuse Neglect 1999*; 23(11):1057–1068.
29. Weitzman BC, Knickman JR, Shinn M. Predictors of shelter use among low-income families: psychiatric history, substance abuse, and victimization. *Am J Public Health* 1992:82(11):1547–1550.
30. Wenzel SL, Koegel P, Gelberg L. Antecedents of physical and sexual victimization among homeless women: a comparison to homeless men. *Am J Community Psychol* 2000; 28(3):367–400.
31. Zlotnick C, Tam T, Roberston MJ. *Adverse Childhood Events, Substance Abuse, and Measures of Affiliation*. Berkeley, CA: Alcohol Research Group, Public Health Institute, 2002.
32. Brabant S, Forsyth CJ, LeBlanc JB. Childhood sexual trauma and substance misuse: a pilot study. *Subst Use Misuse* 1997; 32(10):1417–1431.
33. Harmer ALM, Sanderson J, Mertin P. Influence of negative childhood experiences on psychological functioning, social support, and parenting for mothers recovering from addiction. *Child Abuse Neglect* 1999; 23(5):421–433.
34. Wilsnack SC, Vogeltanz ND, Klassen AD, Harris TR. Childhood sexual abuse and women's substance abuse: national survey findings. *J Stud Alcohol* 1997; 58:264–271.
35. Garrett K. *Field Report for the Study of Alameda County Residents*. Berkeley: Survey Research Center, University of California, 1994.
36. Piazza T, Cheng Y-T. *Sample Design for the Study of Alameda County Residents*. Berkeley, CA: Survey Research Center, University of California, 1992.
37. Zlotnick C, Robertson MJ, Lahiff M. Getting off the streets: economic resources and residential exits from homelessness. *J Community Psychol* 1999; 27(2):209–224.
38. American Psychiatric Association. *DSM-III-R: Diagnostic and Statistical Manual of Mental Disorders*. Washington, DC: American Psychiatric Association, 1987.
39. Anthony JC, Helzer JE. Syndromes of drug abuse and dependence. In: Robins LN, Regier DA, eds. *Psychiatric Disorders in America: The Epidemiologic Catchment Area Study*. New York: The Free Press, 1991:116–154.

40. Office of Employment and Unemployment Statistics. How the Government Measures Unemployment. Report 864; Washington, DC: U.S. Department of Labor, Bureau of Labor Statistics, Division of Labor Force Statistics, 1994.
41. Norusis MJ. *SPSS Advanced Statistics User's Guide.* Chicago: SPSS, Inc., 1990.
42. Muthen LK, Muthen B. *Mplus Users Guide.* Los Angeles: Muthen and Muthen, 1998.
43. Wickens TD. *Multiway Contingency Tables Analysis for the Social Sciences.* Hilldales, NJ: Lawrence Erlbaum Associates, 1989.
44. Newcomb MD. Drug use and intimate relationships among women and men: separating specific from general effects in prospective data using structural equation models. *J Consult Clin Psychol* 1994; 62:463–476.
45. Browne MW, Cudeck R. Alternative ways of assessing model fit. In: Bollen KA, Long JS, eds. *Testing Structural Equation Models.* Newbury Park: Sage, 1993:136–162.
46. Yu C-Y, Muthen B. *Evaluation of Model Fit Indices for Latent Varable Models with Categorical and Continuous Outcomes.* Los Angeles, CA: Social Research Division, Graduate School of Education and Information Studies, UCLA, 2001.
47. Bernstein E, Bernstein J. *Peer educators use motivational intervention to link ED patients to substance abuse treatment.* American Public Health Association Meeting, Boston, MA, November 12–16; 2000.
48. Medrano MA, Zule WA, Hatch J, Desmond DP. Prevalence of childhood trauma in a community sample of substance-abusing women. *Am J Drug Alcohol Abuse* 1999; 25(3):449–462.
49. McDonald WR. *Child Maltreatment 1999.* Washington, DC: U. S. Department of Health and Human Services, Administration on Children, Youth and Families, 2001.
50. McDonald TP, Allen RI, Westerfelt A, Piliavin I. *Assessing the Long-Term Effects of Foster Care: A Research Synthesis.* Madison, WI: Institute for Research on Poverty, 1993.
51. Koegel P, Burnam MA. Alcoholism among homeless adults in the inner city of Los Angeles. *Arch Gen Psychiatry* 1988; 45(11):1011–1018.
52. Calsyn RJ, Morse GA. Correlates of problem drinking among homeless men. *Hosp Community Psychiatry* 1991; 42(7):721–725.
53. Sosin MR, Bruni M. Homelessness and vulnerability among adults with and without alcohol problems. *Subst Use Misuse* 1997; 32(7 and 8):939–968.
54. Bybee D, Mowbray CT, Cohen E. Short versus longer term effectiveness of an outreach program for the homeless mentally ill. *Am J Community Psychol* 1994; 22(2):181–209.

55. Blankertz LE, Cnaan RA, White K, Fox J, Messinger K. Outreach efforts with dually diagnosed homeless persons. *Faro Soc* 1990; 71(7):387–397.

56. Division of Programs for Special Populations. *Case Management for Special Populations.* Washington, DC: Bureau of Primary Health Care, D.H.H.S., 1992.

57. Kirby MW Jr, Braucht GN, Brown E, Krane S, McCann M, VanDeMark N. Dyadic case management as a strategy for prevention of homelessness among chronically debilitated men and women with alcohol and drug dependence. *Alcohol Treat Q* 1999; 17(1/2):53–71.

58. Zlotnick C, Marks L. Case management services among ten federally-funded health care for the homeless projects. *Child Serv Policy Res Pract* 2002; 5(2):113–122.

6

Groups and Organizations

My Culture, My Self (Cultural Differences in Japan and the United States)

Bruce Bower

Cultural differences between Japan and the United States result in unique pressures being placed on members of each society. Americans are encouraged to pursue their individuality, but the Japanese consider themselves to be small members of a larger collective society.

Upon leaving the United States for several months of study at a Japanese university, Leo got a crash course in culture shock. Activities that the undergraduate had enjoyed in his native land, such as playing a match of volleyball with friends, suddenly felt strange and unnatural. Casual volleyball games back home featured a relaxed, cheerful atmosphere and good-humored competitiveness. In Japan, players adopted a grim, no-nonsense manner suited to the application of "ganbaru," a dogged determination to persevere and keep trying until the end of a task.

Leo's Japanese volleyball experience was, to use a culinary analogy, like biting into a cheeseburger and getting a mouthful of sushi. Something about

Japanese life changed the flavor of even the most innocuous items on his menu of customary pursuits. Leo quickly learned to put on a Japanese-style "game face" when he played volleyball, but he just did not feel like himself.

Culture clashes such as this accentuate the fact that largely unspoken, collective assumptions about appropriate social behavior vary greatly from one country or geographic region to another, says Japanese psychologist Shinobu Kitayama of Kyoto University. Moreover, the goals, values, ideas, and behaviors that a person learns and uses as a member of a cultural group have far-reaching effects on mental life, Kitayama argues.

The cherished Western concept of a sovereign self provides a case in point. Consider Leo, whose passport to Japan probably should have been stamped with this brief warning: Bearer comes from culture that treats individuals as independent operators, each of whom must emphasize personal strengths and pump up self-esteem to succeed in life.

In contrast, Japanese culture views individuals as part of an interconnected social web, Kitayama contends. A sense of self develops as a person discerns the expectations of others concerning right and wrong behavior in particular situations. Self-improvement requires an unflagging commitment to confronting one's short-comings and mistakes; their correction fosters harmony in one's family, at work, and in other pivotal social groups.

This cultural perspective appears in various forms throughout East Asia. Its adherents tend to write off the European-American pursuit of self-esteem as an immature disregard for the relationships that nurture self-identity, Kitayama says.

A growing body of research explores the ways in which cultural perspectives, such as an emphasis on personal independence or social interdependence, shape psychological tendencies, as seen in strivings for self-enhancement or self-criticism. Proponents of this approach, who call themselves cultural psychologists, argue that many seemingly natural mental tendencies detected by research in Western nations vanish or change drastically in other cultures.

"The cultural underpinnings of self-enhancement and other mental phenomena have eluded the attention of many North American researchers," Kitayama contends. "Humans have evolved to live a social life in groups. By arranging social life in different ways, cultures affect psychological processes."

Richard A. Shweder, a psychologist at the University of Chicago, sees culture as a prominent shaper of minds. A culture, in his view, serves as a flexible learning system that transforms basic biological capacities into meaningful thoughts and behaviors shared by its members. Even disillusioned souls who rebel against the social norm make a point of thumbing their noses only at their own cultural practices.

In Japan, Shweder notes, parents and teachers encourage youngsters from infancy onward to seek out other people and to modify their behavior according to social rules and expectations. The importance of adjusting to social feedback is expressed in the Japanese word for self, "jibun," which means "my share of the shared space between us."

Self-improvement thus hinges on finding and repairing personal failures to advance the social enterprise rather than basking in one's unique advantages over others. For the Japanese, Shweder says, it is usually a relief to know that one is "hitonami," average as a person; in the United States, people who achieve this insight run in anguish to psychotherapists and self-help gurus.

Two studies published in the June *Journal of Personality and Social Psychology* delve into the different manifestations of self that arise in Japan and the United States.

One investigation, directed by Kitayama, identifies a penchant for self-enhancement in U.S. college students and an affinity for self-criticism in Japanese college students. The researchers first asked 90 Japanese and 65 U.S. students to describe as many situations as possible in which their own self-esteem either increased or decreased. From these responses, the researchers developed a list of 400 situations, half from each country. The number of esteem-enhancing, or success, incidents on the list equaled the number of esteem depressors, or failure situations.

A second set of U.S. students then read the list and rated more of the success than of the failure situations as likely to affect their own self-esteem. They also said that their self-esteem rose more sharply after successes than it dropped after failures. In contrast, a second group of Japanese students reported tendencies to focus on failure situations when evaluating their own self-esteem and to modulate self-esteem more strongly after experiencing a failure.

Japanese-style self-criticism appeared in a slightly weaker form in a group of Japanese students temporarily studying at a U.S. university. Self-critical attitudes inculcated in Japan appear largely immune to brief forays into Western culture, the scientists hold.

The same cross-cultural disparity emerged when Japanese and U.S. students read the fist of 400 situations and described how each incident would influence the self-esteem of a typical undergraduate at their respective schools. Failure situations loomed large for the Japanese, whereas Americans focused on successes that would boost the confidence and satisfaction of their academic compatriot.

The ready assignment of self-criticism to a hypothetical student by Japanese volunteers suggests that they regard it as a completely natural attitude, akin to a guiding sense of humility, Kitayama argues. This observation counters the view of researchers who contend that the Japanese criticize themselves in a self-deprecating way intended to mask their inner desire for personal success.

North American college students also emphasize the positive qualities of groups to which they belong far more than their Japanese counterparts do, the second investigation finds. Asked to evaluate both a close family member and their own schools, students at two top-ranked Canadian universities gave more favorable ratings than did students attending two highly regarded Japanese universities, report Steven J. Heine of the University of Pennsylvania in Philadelphia and Darrin R. Lehman of the University of British Columbia in Vancouver.

Volunteers selected a family member and then rated the extent to which he or she possessed 10 traits, including attractiveness, intelligence, cooperativeness, and dependability. They also ranked characteristics of their own educational institution, such as its academic reputation and the accomplishments of its graduates, as well as attributes of a typical student at the school.

Although the Japanese people report pride and happiness at being associated with a top-flight school or other social unit, they refrain from making unrealistically positive appraisals of their groups, Heine asserts.

"Japanese do not just say that they are no better than average, they truly seem to believe it," he says.

A related study by Heine and Lehman, now submitted for publication, indicates that the personal goals to which individuals aspire are further out of reach for Japanese students than for Canadian students, at least according to their own accounts. Even so, the Canadians more often report feeling depressed and dissatisfied about their failures to measure up to personal ideals.

"People have a need to view themselves as good and meaningful citizens of their cultures," Heine remarks. "Thinking highly of yourself does not seem to be particularly important in Japan, where people want to secure a sense of belonging to social groups and ensure that others are satisfied with their contribution to those groups."

Cultural psychologists hope to do more than challenge Western psychology's focus on self-esteem. For example, based on other recent experiments submitted for publication, Kitayama and Kyoto University colleague Takahiko Masuda argue that Japanese volunteers often do not assume that another person's behavior corresponds to his or her private attitudes.

A "correspondence bias" that equates behavior with internal attitudes or personality traits appears in numerous studies conducted in Western countries. Researchers have theorized that it demands more of a social thinker to consider the situational influences on another's behavior than simply to assume that behavior reflects an internal attitude.

In one classic U.S. study, participants assumed that the writer of an essay arguing in favor of Fidel Castro's regime held fairly strong pro-Castro views. It mattered little whether they had been told that the writer had chosen to express pro-Castro views or had done so on instruction by a professor.

Japanese college students pay close attention to information about social inducements to behavior, such as a professor's instructions to express a certain political opinion, Kitayama and Masuda find. Japanese volunteers expressed a correspondence bias only if they had no access to information about social influences on the essay writer.

Japanese participants also tended to assume that other people's attitudes about a sensitive political issue (keeping or closing a U.S. military base on Okinawa) might differ from their own. This contrasts with data showing that U.S. volunteers generally believe that others agree with their views.

Moreover, in the April *Personality and Social Psychology Bulletin*, Heine and Lehman reported that U.S., but not Japanese, students substantially changed their opinions of pieces of music in an attempt to imitate more socially competent people. The U.S. students revised their ratings of popular compact disc recordings after being told that they scored lower than most of their peers on an inventory of positive personal attributes. In this situation, the Japanese students felt no need to qualify their musical attitudes.

U.S. students felt psychologically threatened by evidence of their personal shortcomings, as assessed on the inventory, and changed their musical ratings in the hope of demonstrating a similarity to high-scoring peers, the researchers hold.

This kind of response to inconsistencies between personal attitudes and behaviors, a phenomenon known as cognitive dissonance, has been studied in North America for 40 years. Further work will address the subtleties of cognitive dissonance and other psychological phenomena across East Asian cultures, Heine and Lehman say.

Cultural psychologists' skepticism about the universality of mental responses studied in the Western world has garnered mixed reviews. Some scientists, such as psychologist Daniel T. Gilbert of Harvard University, view the results to date as too preliminary to challenge the cross-cultural existence of the correspondence bias and other much-studied effects.

"There have got to be some cultural differences in what psychologists study, but much more research is needed to prove the fascinating hypotheses of cultural psychologists," Gilbert remarks.

Phoebe C. Ellsworth of the University of Michigan in Ann Arbor sees a deeper lesson in the studies conducted so far. "We've been awfully provincial in assuming that our findings in social psychology reflect human nature," she argues.

In unpublished research related to the Japanese results, Ellsworth and a colleague find that Chinese undergraduates have no difficulty in describing how a group of cartoon fish feels in various situations, such as assuming that the group is angry when one fish departs. Their U.S. counterparts generally get flustered or confused by this task and often want to know which individual fish they're being asked about.

Despite a shared emphasis in China and Japan on the social interdependence of individuals, cultures in these two countries probably differ in ways that markedly influence how people view themselves, Ellsworth adds.

Individuals everywhere maintain a sense of self on three levels, theorizes Marilynn B. Brewer of Ohio State University in Columbus. These consist of a personal identity, identities tied to membership in various groups, and social identities that arise when interacting with others. For example, a woman assumes an identity as a mother when she is out in public with her children.

Different cultures encourage people to develop distinctive blends of these identities, Brewer suggests.

"What's important about research such as Kitayama's is that it implies that cultures create situations that support the definition of the self," she says.

7

Deviance and Crime

The Effect of Video Games on Feelings of Aggression

Derek Scott

Fueled by the media, the controversy over whether playing popular arcade/computer games increases aggressiveness has only been compounded by inconsistencies within empirical research. This experiment, conducted with university students in Scotland, was designed to explore some of these inconsistencies. Aggressiveness was manipulated as the independent variable. As dependent variables, the Buss-Durkee Hostility Inventory (Buss & Durkee, 1957) and the Eysenck Personality Questionnaire (EPQ; Eysenck & Eysenck, 1975) were used. There was no linear pattern in aggressive affect change across three games that contained varying levels of violence. Results are discussed in terms of the general lack of support for the commonly held view that playing aggressive computer games causes an individual to feel more aggressive.

The link between television viewing and violence has been researched and debated for some time (Andison, 1977; Berkowitz, 1984; Eron, 1982; Gunther, 1981; Pearl, Bouthilet, & Lazar, 1982). More recent concerns have included how not only television but also cinema and video viewing might influence levels of aggression (Screen Violence, 1993): "Over the last 10–15 years, the limited data suggest, if anything, a decrease in the quantity of violence on the four main TV channels, although information on shifts in the type of violence is lacking" (p. 353).

During the last decade, attention and accusations within the media have turned more to the meteoric rise in popularity of arcade-type home computer and console games. Considerable anecdotal evidence abounds about how teenagers are affected by shoot-em-up and beat-em-up games. Zimbardo (1982) remarked that video games are so addictive to young people that they may be socially isolating and may actually encourage violence between people. Another comment came from the surgeon-general of the United States, who expressed his personal view that video games were one of the root causes of family violence in America. He was quoted as saying that children "are into the games, body and soul—everything is zapping the enemy. Children get to the point where when they see another child being molested by a third child, they just sit back" (Koop, 1982).

The similarities between television and video games have also been noted (Silvern & Williamson, 1987). Both have entertainment value, violent content, and various physical feature similarities (action, pace, and visual change). A majority of video games are violent in nature and feature death and destruction (Dominick, 1984; Loftus & Loftus, 1983). In the survey by Bowman and Rotter (1983), 85% of the video games examined (n = 28) involved participation in acts of simulated destruction, killing, or violence. In addition, concern has been raised that video games may have a greater adverse effect than television because of the active involvement of the player (Bowman & Rotter; Greenfield, 1984). This issue is further detailed by Griffiths (1991).

Because most research into television violence does demonstrate a relationship between the exposure to aggression and subsequently exhibited aggression, investigations of the effects of video game playing usually have predicted a similar relationship. However, many variables are involved, and researchers offer no clear statement on the role of game playing and aggressiveness. Parameters include, for instance, gender, age grouping, expressed hostility (feelings of aggressiveness) versus exhibited aggression (overt behavior), the behavioral measurement (e.g., toward a life-size doll, or in terms of shocks administered from an "aggression machine"), experimental duration of exposure (time spent playing), and personality traits. Also, studies may be laboratory based or observational.

Findings from studies using various subject groups and various methodologies have been mixed. Dominick's (1984) questionnaire-based study reported a significant relationship between video game playing and aggressive delinquency in adolescents. Conversely, in another questionnaire study (also involving teenagers), Kestenbaum and Weinstein (1985) reported that aggressive games had a calming effect.

Researchers have compared children's free-play behavior after aggressive and nonaggressive video game play (Cooper & Mackie, 1986; Schutte, Malouff, Post-Gorden, & Rodasta, 1988; Silvern, Williamson, & Countermine, 1983). In laboratory research, Cooper and Mackie found that girls increased their aggressive free play after an aggressive game and increased quiet play after a nonaggressive game. The free play of the boys, however, was not significantly altered by either game.

Using a similar paradigm with younger children, Schutte et al. (1988) evaluated the changes in 5- to 7-year-old children after they had played violent or nonviolent video games. In subsequent free play, the children involved in the aggressive game were more violent. In an observational study of free play, Silvern and Williamson (1987) demonstrated increased aggression and decreased prosocial behavior in 4- to 6-year-olds after playing violent video games. Dominick (1984) used a video game survey, and Nelson and Carlson (1985) examined the type of video game preferred in a free-choice situation.

In addition to the problems of inconsistency, certain studies seem to have been methodologically flawed. Graybill, Kirsch, and Esselman (1985) found that children who played the violent video game exhibited fewer defensive fantasies and more assertive fantasies than did the children who played the nonviolent game. In addition, they noted that the barrier responsible for frustration was more salient for the nonaggressive girls after frustration than for the aggressive girls. They concluded that playing the violent video game may have had some short-term beneficial effects for the children, but they later acknowledged that the projective technique used (the Rosenzweig Picture-Frustration Study) was not a valid measure (Graybill, Strawniak, Hunter, & O'Leary, 1987).

In this second study of Graybill et al. (1987), children of the same age group played a violent or nonviolent video game for just 7 min (and observed a partner play for 7 min). Despite using projective, behavioral, and two self-report measures, this study again showed no differences between the violent and nonviolent conditions. The behavioral measure required the children to press buttons that ostensibly would hurt another child.

Apart from the ethical problems involved, it is noteworthy that one of the three nonaggressive games used consisted of a frog catching and devouring flies with its tongue. Another was a Pac Man type game where a mouth chases food items and has to gobble these up before being itself destroyed by other chasing mouths. These do not seem good examples of a nonaggressive games. Indeed, Cooper and Mackie (1986) reported that the girls in their study saw little difference in aggressiveness between Pac Man and their aggressive game, Missile Command.

Graybill et al. (1987) suggested that graphics are not as realistic in video games as on television, but since then technological changes have led to remarkably realistic graphics. Graybill et al. (1987) also argued that the differences in typical results for TV viewing and their video game study were possibly due to their subjects' being in the presence of a peer partner, and that the presence of a peer may have caused the children to be more attentive to the scores than to the content of the games.

Winkel, Novak, and Hopson (1987) found no relationship in adolescents between playing a violent video game and aggression, here defined as the amount of money one subject fined another subject (actually a computer) from a total payment. Winkel et al. suggested that, in the adolescent age group, personality traits and social contextual variables were more important as determinants of behavior than was exposure to video games. In a somewhat older age group, Anderson and Ford's (1986) questionnaire study demonstrated that violent video games increased hostility among university undergraduates.

In brief, there have been inadequacies and inconsistencies in choice of both independent and dependent variables. It is therefore necessary to attempt a more valid and thorough assessment of the possible associations between video game playing and aggression. What actually constitutes aggression and how it may be quantified has been the subject of much debate. For example, in their treatise on the measurement of aggression, Edmunds and Kendrick (1980) stated that aggression may more precisely be classified into aggression, which would seem generally to cover overt and direct behaviors, and aggressiveness, which is typically represented by hostile feelings.

One major problem with previous studies is that they contain no single, standardized and well-validated measure of aggressiveness that identifies its various types. Although several MMPI-derived inventories measuring aggression/hostility have been developed since the mid-1950s and prior to the Buss-Durkee Inventory (1957), no questionnaire was available that gave more than a very global measurement. For example, a nonverbal, physically assaultive individual might receive the same score as a nonassaultive, verbally aggressive person.

The Buss-Durkee Inventory (1957) groups items into subscales representing various aspects of aggression and hostility, thus provides a finer analysis of the general concept of aggression, and classifies seven types of aggressiveness (in the sense of reported feelings, rather than demonstrated behaviors) derived from factor analysis studies. Often, as in the present study, the Guilt scale is omitted, which leaves 66 items classified into the following subscales:

1. Assault—physical violence against others. This includes getting into fights with others but not destroying objects.
2. Indirect Hostility—both roundabout and undirected aggression. Roundabout behavior such as malicious gossip or practical jokes is indirect in the sense that the hated person is not attacked directly but by devious means. Undirected aggression, such as temper tantrums and slamming doors, consists of a discharge of negative affect against no one in particular; it is a diffuse rage reaction that has no direction.
3. Irritability—a readiness to explode with negative affect at the slightest provocation. This includes quick temper, grouchiness, exasperation, and rudeness.
4. Negativism—oppositional behavior, usually directed against authority. This involves a refusal to cooperate that may vary from passive compliance to open rebellion against rules or conventions.
5. Resentment—jealousy and hatred of others. This refers to a feeling of anger at the world over real or fantasied mistreatment.
6. Suspicion—projection of hostility onto others. This varies from merely being distrustful and wary of people to beliefs that others are being derogatory or are planning harm.

7. Verbal Hostility—negative affect expressed in both the style and content of speech. Style includes arguing, shouting, and screaming; content includes threats, curses, and being overcritical.

It seems reasonable to assume that the inconsistency of previous findings is due not only to inadequate consideration of what constitutes aggressiveness but also to personality differences. Individuals of differing personality types undoubtedly react differently to particular situations and events. Of the numerous personality assessments available, one of the most reliable and widely validated is the Eysenck Personality Questionnaire (EPQ; Eysenck & Eysenck, 1975), which consists of four scales: E (introversion-extraversion), N (stability-instability), P (tough-mindedness), and L (social desirability, or the Lie scale).

My principal aim in the present study was to investigate to what extent, if any, aggressive computer game playing would have on individuals of differing personality composition and in which particular aspects of aggressiveness this might be experienced. The study was limited to measuring aggressiveness, or aggressive affect, rather than overt aggression. It was also intended to examine the different effects of exposure in male and female participants. To avoid confounding effects of age, educational level, and so forth, I used a homogeneous group of university students.

The next issue was choice of stimulus material—the games played. I decided to have three levels of aggression or violence in the content of the games; nonaggressive, moderately aggressive, and highly aggressive. I hypothesized that there would be a linear increase in aggressive affect after playing nonaggressive, moderately aggressive, and highly aggressive games. I also examined interactions among gender and aggressiveness, among gender, aggression level, and personality, and among gender, aggression level, and type of aggressiveness.

METHOD

Participants and Equipment

A total of 117 students from Strathclyde University participated (42 men, 75 women).

The main piece of equipment was an Amstrad "Mega" PC486SLC, which incorporates a "Sega" compatible games cartridge drive. The monitor was a VGA color Amstrad PC14DSM 14[inches] designed for the domestic games market. The computer contained an Ad-Lib compatible sound card, and the monitor contained built-in stereo speakers. The manual peripherals involved were an "ergonomically designed" Quick Shot QS-123 ("Warrior 5") analog joystick featuring fire and auto-fire controls and a two-handed Amstrad Mega PC games paddle.

The nonaggressive game was Tetrisc, a Shareware version of Tetris. Essentially, this game involves manipulating geometric blocks as they fall down the screen. It contains the joystick movements and fire button involvement, sound (musical accompaniment changing along with level of game play achieved), color, and the necessarily fast speed of control that are fundamental features of the other two games; yet Tetrisc had no aggressive element.

The moderately aggressive game was Overkill (Shareware), a typical, modern arcade-type "space blasters," joystick controlled, vertically scrolling game. One has to shoot up as many alien ships as possible and maneuver to avoid being hit. It is accompanied by firing sound effects and digitized cries of "kill kill kill" when hits are made.

The highly aggressive game was a Sega Mega Drive cartridge of Fatal Fury. It is generally regarded as one of the most violent paddle games available outside of the amusement arcades. Essentially, the player takes the form of a martial arts expert and has to kick, punch, head-butt, and so on, the (computer-controlled) opponents before they do likewise to oneself. The graphics are large-scale, realistic-looking human characters. Attention has undeniably been paid by the programmers to convey an impression of pain and injury. Sounds of thumps and groans accompany the bodily impacts.

Procedure

A before-and-after between-subjects design was used, in which each student participated in only one condition (played one game). An equal number of men and women took part in the three conditions. Split-half versions were created for the EPQ and Buss-Durkee inventories, with an occasional repetition across versions to accommodate an uneven number of items on some subscales. Half the subjects were administered Version A first, and half were given Version B first.

The students were told that the study concerned a hand-eye coordination task in relation to personality. They were asked to complete one half of the (merged) inventories and were introduced to the games with standardized instructions. They played the game for 10 min, after which they completed the second half of the inventory. They were then asked briefly about previous experience and present involvement with this sort of entertainment. Before being debriefed, many of the subjects were asked to rate the game in terms of aggressive content on a 0–10 scale.

Results

The results of 3 participants were discarded because of an overhigh score (> 14) on the combined two parts of the L (social desirability) subscale of the EPQ. Replacement data were obtained from 3 other participants.

Aggression ratings were derived as a change in aggressiveness, measured before and after participation; they could be positive (more aggressiveness) or negative (less aggressiveness). As there were unequal numbers of items in the various subscales of the Buss-Durkee Inventory, these data were calculated as percentages and percentage changes rather than as absolute values.

Table 1 contains the means of the total aggressiveness change between the levels of aggressiveness and between men and women. Table 2 contains the results on the aggressiveness subscales.

A three-way analysis of variance (ANOVA) with two between factors (gender and level of aggression) and one within factor (type of aggression) revealed a significant difference, $F(2, 111) = 4.39$, $p < .05$, only between levels of aggression. There were no second- or third-order interactions.

This difference between levels of aggression was clearly between the nonaggressive versus moderately aggressive games, and between the highly aggressive versus moderately aggressive games rather than between nonaggressive games and highly aggressive games, and hence warranted no ad hoc statistical analysis. In other words, there was no linear trend.

TABLE 1. Means of Total Aggressiveness Change (in Percentages) Between Levels of Aggressiveness and Between Men and Women Participants

	Before	After
NONAGGRESSIVE		
Men	35.357	67.436
Women	−2.720	70.574
Total	10.949	76.794
MODERATELY AGGRESSIVE		
Men	−25.286	84.063
Women	−33.720	76.804
Total	−30.720	78.485
HIGHLY AGGRESSIVE		
Men	−20.714	75.667
Women	0.154	82.185
Total	11.840	84.833

TABLE 2. Means and Standard Deviations on the Aggressiveness Subscales

Subscale	M	SD	Maximum	Minimum
Assault	1.97	21.79	60.00	−40.00
Indirect Hostility	2.03	25.17	60.00	−60.00
Irritability	−3.26	24.64	50.00	−67.00
Negativism	−0.32	35.74	77.00	−100.00
Resentment	3.13	21.14	57.00	−48.00
Suspicion	0.47	30.77	80.00	−80.00
Verbal Hostility	3.13	21.14	57.00	−48.00
Total aggression	−6.56	80.53	250.00	−170.00

A Kruskal-Wallis one-way ANOVA showed no significant differences between overall change in aggressiveness (total) scores and level of game aggression. Nor was there any significant difference (point-biserial correlation) between total aggressiveness change and game aggression level.

Pearson's correlation coefficients were computed on total aggressiveness scores and also on the seven subscales versus the personality variables of extraversion, neuroticism, and psychopathy (a total of 24 correlations); all failed to show significant differences. The mean average (and standard deviations) of aggressiveness content rating given by subjects on the nonaggressive, moderately aggressive, and highly aggressive game were 0.8 (1.4), 5.7 (2.4), and 6.7 (1.8), respectively. Figure 1 shows mean before and after levels of total aggressiveness across the three variables for men and women. Finally, it must be remarked that there was a distinct irregularity across the results.

Discussion

I had hypothesized that there would be a linear increase in aggressive affect after playing nonaggressive, moderately aggressive, and highly aggressive games, but no such increase occurred. The overall pattern was that the moderately aggressive game substantially decreased feelings of aggression, whereas the highly aggressive game resulted in much less of an increase in aggressiveness than I expected, although no more so than occurred in the control game. Generally, the participants did regard the games to be more aggressive in the expected order, although the difference of feelings about the two aggressive games was not as great as might have been expected. However, this pattern could not account for the irregularity of aggressiveness changes that were found.

The greatest change was among the men who participated in the nonaggressive game. They showed substantially more overall aggressiveness afterward. However, the men who played the nonaggressive game had generally been considerably (although at chance level) less aggressive both before and after playing than the men randomly assigned to the two aggressive games. This finding only emphasizes the fact that individual variability is more important than variability in affect induced by playing computer games.

I also examined second- and third-order interactions between gender, aggression level, and personality. Despite careful choice of variables and materials, none were found.

For over a decade, proponents of video games have championed their educational value (Malone, 1981), value for social interaction and growth, and therapeutic value (Leerhsen, Zabarsky, & McDonald, 1983). However, an explanation for the present results may be more physiological than psychological. One approach to understanding the causes of aggression emphasizes the role played by the sympathetic nervous system, with heightened sympathetic activity seemingly facilitating overt aggression. Winkel et al. (1987) found that, for male adolescents, personality traits and heart rate were separately related to aggression. Personality characteristics similar to those of the Type A individuals were related in a positive direction to heart rate in women. Winkel et al. concluded that there was no evidence that the link between game playing and aggression is due to mimickry per se. Their results suggest that home video games, regardless of their aggressive content, may stimulate a more violent reaction in girls than in boys.

The interactions between the variables are obviously complex, and glib statements relating aggression to game playing, whether appearing in the mass media or in scientific journals, seem totally unwarranted. In addition, Cooper and Mackie (1986) suggested that only their female participants felt there was little difference in aggressiveness content between the games played—Pac Man and Missile Command.

In general, one should not overgeneralize the negative side of computer games playing. During the last decade, the market for fun-laden educational software has exponentially increased. Also, Funk (1992) suggests that playing home video games may have a less adverse impact on academic functioning than playing in an arcade.

There may also be individual differences in the effect of game playing. Some people may be able to spend a great deal of their free time playing arcade videos without any resulting aggression. Huesmann (1982) concluded that children who are exposed to the least violence may be the most aroused and most likely to act aggressively. Because girls are likely to have less experience with violence, they are the group likely to be more aroused by the exposure.

The present study points up the need for considering the strength of individual differences when researching the effects of video games on feelings of aggression.

REFERENCES

Anderson, C. A., & Ford, C. M. (1986). Affect of the game player: Short term effects of highly and mildly aggressive video games. *Personality and Social Psychology Bulletin, 12,* 390–402.

Andison, F. S. (1977). T.V. violence and viewer aggression: A cumulation of study results 1956–1976. *Public Opinion Quarterly, 41,* 314–331.

Berkowitz, L. (1984). Some effects of thoughts on anti- and prosocial influences of media events: A cognitive-neoassociation analysis. *Psychological Bulletin, 95,* 410–427.

Bowman, R. P., & Rotter, J. C. (1983). Computer games: Friend or foe? *Elementary School Guidance and Counselling, 18,* 25–34.

Buss, A. H., & Durkee, A. (1957). An inventory for assessing different kinds of hostility. *Journal of Consulting Psychology, 21,* 343–349.

Cooper, J., & Mackie, D. (1986). Video games and aggression in children. *Journal of Applied Social Psychology, 16,* 726–744.

Dominick, J. R. (1984). Videogames, television violence, and aggression in teenagers. *Journal of Communication, 34,* 136–147.

Edmunds, G., & Kendrick, D. C. (1980). *The measurement of human aggressiveness.* Chichester: Ellis Horwood.

Eron, L. D. (1982). Parent-child interaction, television violence, and aggression of children. *American Psychologist, 37,* 197–211.

Eysenck, H. J., & Eysenck, S. B. G. (1975). *Manual of the Eysenck Personality Questionnaire.* London: Hodder & Stoughton.

Funk, J. B. (1992). Video games: Benign or malignant? *Developmental and Behavioral Pediatrics, 13,* 53–54.

Graybill, D., Kirsch, J. R., & Esselman, E. D. (1985). Effects of playing violent versus nonviolent video games on the aggressive ideation of aggressive and nonaggressive children. *Child Study Journal 15,* 199–205.

Graybill, D., Strawniak, M., Hunter, T., & O'Leary, M. (1987). Effects of playing versus observing violent versus nonviolent video games on children's aggression. *Psychology: A Quarterly Journal of Human Behavior, 24,* 1–8.

Greenfield, P. (1984). *Media and the mind of a child: From print to television, video games and computers.* Cambridge: Harvard University Press.

Griffiths, M. D. (1991). Amusement machine playing in childhood and adolescence: A comparative analysis of video games and fruit machines. *Journal of Adolescence, 14,* 53–73.

Gunther, B. (1981). Measuring television violence: A review and suggestions for a new analytical perspective. *Current Psychological Reviews, 1,* 91–112.

Huesmann, L. R. (1982). Video games and aggression. In D. Pearl, L. Bouthilet, & J. Lazar, (Eds.), *Television and behavior: Ten years of progress and implications for the eighties (Vol. 2). Technical reviews.* Washington, DC: U.S. Government Printing Office.

Kestenbaum, G. I., & Weinstein, L. (1985). Personality, psychopathology, and developmental issues in male adolescent video game use. *Journal of the American Academy of Child Psychiatry, 24,* 325–337.

Koop, E. (1982). Surgeon general sees danger in video games. *New York Times,* November 10th, p. A16.

Leerhsen, C., Zabarsky, M., & McDonald, D. (1983). Video games zap Harvard. *Newsweek, 101* (July), p. 92.

Loftus, G. A., & Loftus, E. F. (1983). *Mind at play: The psychology of video games.* New York: Basic Books.

Malone, T. W. (1981). Toward a theory of intrinsically motivating instruction. *Cognitive Science, 4,* 333–370.

Nelson, T. M., & Carlson, D. R. (1985). Determining factors in choice arcade games and their consequences upon young male players. *Journal of Applied Social Psychology, 15,* 124–139.

Pearl, D., Bouthilet, L., & Lazar, J. (1982). *Television and behavior: Ten years of progress and implications for the eighties (Vol 1). Summary report.* Washington, DC: U.S. Government Printing Office.

Schutte, N. S., Malouff, J. M., Post-Gorden, J. C., & Rodasta, A. L. (1988). Effects of playing video games on children's and other behaviors. *Journal of Applied Social Psychology, 18,* 454–460.

Screen violence. (1993, August). *The Psychologist,* 353–356.

Silvern, S. B., & Williamson, P. A. (1987). The effects of video game play on young children's aggression, fantasy, and prosocial behavior. *Journal of Applied Developmental Psychology, 8,* 453–462.

Silvern, S. B., Williamson, P. A., & Countermine, T. A. (1983). Aggression in young children and video game play. Paper presented at the biennial meeting of the Society for Research in Child Development, Detroit, April, 1993.

Winkel, M., Novak, D. M., & Hopson, H. (1987). Personality factors, subject gender, and the effects of aggressive video games on aggression in adolescents. *Journal of Research in Personality, 21,* 211–223.

Zimbardo, P. (1982). Understanding psychological man: A state of the science report. *Psychology Today, 16,* 15.

8

Global Stratification

Value-Adding Information: Virtual Conferencing, a Telecommunication Pathway to the Future

Joan Edgecumbe

A virtual professional health event in Australia involving international telecommunication conferences and the Internet illustrated the health care communications style of the future. The event highlighted education for health care providers and innovations in information technology. The global exchange of information offered participants an opportunity to recognize the vast amount of medical information global communication can provide.

The information or digital age has turned information communication upside down and inside out. This article describes use of the Internet and World Wide Web information flow and communication for a virtual professional health national event, in Australia. The stored information and subsequent dialogue were available nationally and internationally, making the telecommunication conference a global one. Defined are the background, vision, and assumptions on which the conceptual conference structure was ultimately designed, implemented, and evaluated, including the implications for value-adding nursing information. Key words: Internet, communication, computer, conference, World Wide Web

Australia is a big country, slightly larger in geographic size than the United States of America, minus Alaska. The Australian community is spread mainly from coastal to rural towns and cities around the continent's seaboard. Like many large continents, rural and remote health professionals do not have the opportunity to attend conferences held around their country, predominantly due to time availability, cost, and staff replacement. It is also becoming more difficult for interested professionals, especially students, to afford all the direct and indirect costs, time, and convenience associated with attendance at multiple professional events. The use of telecommunication tools reduces barriers to professionals attending conferences[1] and communication groups. Earlier attempts to use telecommunication technology were constrained by the limits of provider infrastructure and client access to services. Anderson, in his article "History of the Virtual Conference," a resource article found on the World Wide Web (WWW), describes concisely the history, funding, learning activities, tools, and definition of a virtual conference from early days in 1990 to today.[2] It was with some relief that Anderson's article was found on the WWW as a resource, but only after this virtual event had taken place. For the purpose of this article the virtual conference will be referred to as the event.

VISION

The event was envisioned to facilitate distant telecommunications conference attendance (connectivity) and ongoing education for health professionals, using the means of the information superhighway. Many statements have been made by professionals within Australia, epitomized by an anonymous e-mail communication from a Health Informatics Conference (HIC) '95 virtual attendee, highlighting the ongoing education needs of rural and remote health workers and their inability to attend on-site events due to the many prohibitive factors they experience (geography and work release time). These factors determined the vision for this event. Today's technology has the capacity for value-adding information and allowing connectivity through communication between health professionals both within Australia and internationally. The assumption was that with the availability of scholarship material, in the form of conference paper abstracts and telecommunications information flow, education could be derived from the exchange of participants' knowledge with continued dialogue. Structuring an event and making it available via a virtual information superhighway conference was the primary hurdle. The limiting factors were predetermined as the conversion and storage of required information on a WWW site, education regarding the event, and the Internet access availability to prospective attendees.

HISTORY

The sponsoring organization, the Health Informatics Society of Australia (HISA), is a small, not-for-profit professional association without access to the necessary funding, skills, or resources required for the development and implementation of such an event. Furthermore, it was not known if this type of event had indeed ever occurred before. The only known success factors were those of Internet list servers and bulletin boards that are positioned on computer servers around the world. To further the vision and initial development of the event, a business and industry alliance partner was perceived as necessary for the concept to progress. An alliance was formed with a specialist private Internet medical supplier who currently provides Internet, WWW communication, and education services to medical practitioner groups within the country. Both organizations had until this time never conceived the notion of developing a virtual conference or a communication event of any type; therefore, development time and costs to both parties were unknown.

ORGANIZATION

Prior to the initial development phase, financial and participatory business decisions were required by both parties. These decisions involved the developer investing resource time and provider costs in exchange for conference design, information, general administration, and access to all conference scholarship from the society for its existing medical customers as well as prospective event attendees. As the objective was education and communication connectivity and the event style deemed to be leading-edge in design, an important consideration by both parties was the availability of telecommunication education for attendees. Hands-on education at the conference site for novice or beginner registrants to participate was a necessity; however, education for the Internet only attendee was at the time not possible. A second business consideration regarded the marketing and advertising of the event that would highlight its existence and define advantages of all attendees. The two developing partners agreed that education needed to be both on site within the conference exhibition for access by any on-site attendee and within the structural design of the event for the Internet-only attendee. Thus, a sound structure was designed and an on-site telecommunication center was planned, at a cost to the conference. Three strategically placed free-standing computers were connected by a local telephone line (ISDN = 1 microlink) to the provider's local Internet service site logged onto the HIC'95 virtual event. The computer terminals remained online for the duration of the face-to-face event, making delegate learning and access continually available. A distant formal Internet attendee education program could not be managed; however, a virtual attendee had only to request assistance on any matter and another attendee would respond to the call.

IMPLEMENTING THE WWW SITE AND ELECTRONIC CONNECTIVITY

The structure of the telecommunication storage WWW site (or Web pages) was defined by the initiating society, HISA, and situated on a WWW server in the far north of the country, at Brisbane, Queensland. A WWW site is a series of pages of information, coded in hypertext markup language (HTML) linked in a meaningful way that enables the investigator to navigate from and to each individual page, like an electronic book. The basis for the structure or design was the previously developed hard-copy conference provisional program. The already printed provisional program in its current format ideally suited the WWW site structure and design, including the ease of use and content questions. Content included a description of the society, HISA, the general conference information, and conference sessions, with scholarship added from accepted contributing authors' abstracts, site drawings, and general registration. Another section of WWW content, or pages, consisted of general information about the virtual event, external Internet registration, and the administration. The event was described in session date and time sequence with the abstracts linked to the appropriate conference session. The exhibition and industry attendees were also listed, along with an exhibition floor plan that included the conference sessions plan. The WWW site was built to vision, objectives, expectation, and plan.

At the commencement of the project and during the developmental phase of the event, the title HIC'95 Virtual Conference was listed on the Internet "new resources" registers and colleagues from around the world were notified via personal Internet messages. The virtual conference was designed as a "closed" area, secured by the allocation and use of individual passwords, allowing only registered conference event attendees to read and download the event abstracts. This area also housed the multiple Internet dialogue areas for virtual conversation and communication. The Internet "surfing attendee" selected admission to the virtual conference area via the electronic registration form housed in the information area. Once registration was completed by the prospective attendee and posted via the Internet to the provider and the host society, then an Internet posting of the electronic registration form generated a computer password from the hostserver, which was e-mailed to the administrator for the official registration of the virtual attendee. A return e-mail notification to the attendee of his other individual password and request for a minimal registration fee enabled entrance to the secure virtual conference area of the WWW site for the duration of its being. Virtual attendees were able to visit the virtual conference whenever they wished from wherever they were and to participate in dialogue that was generated from the abstracts, their knowledge, and fellow attendees. A record of all attendees' Internet addresses was maintained by the society administrator for further inquiry purposes. Inquiry was usually loss of, or difficulty with, individual passwords and sometimes answering mail regarding difficulty gaining entry to the virtual conference section.

The implementation phase of going live with abstract information and dialogue windows was the most difficult, time consuming, and costly component for the development team, plus the most exciting. Implementation difficulties were mainly due to the skills and time required for text conversion for the number of abstracts and general information. The excitement for the developing team was the reality of the WWW event for participants following testing and measurement against WWW construction standards.

An electronic virtual welcome was placed within each conference stream or session, understanding some attendees would discriminate in their virtual attendance and only go to topic-specific sessions, while others would surf the entire event by every session. To coincide with "going-live," the administrator conscripted fellow society members who were Internet proficient to assist with monitoring and moderating the event. Each moderator was allocated a virtual stream or session known to be within the moderator's knowledge, interest, and comfort zone. The moderators' responsibilities were, when time permitted, simply to monitor the particular allocated session and Internet activity, direct attendees to other electronic sites or persons for further information, and sometimes inject the electronic conversations.

IMPLEMENTATION OBSTACLES

Internet security, ease of use, and navigation were highlighted as design standards requiring attention at the commencement of the project by both developing partners. Poor or incomplete construction in these areas would pose obstacles to the success of the event. With hindsight knowledge, the administration task of logging the virtual attendees and maintaining their password access and individual addresses proved the most difficult to manage in time allocation and work intensity. While formal registration was considered extra work for the administrator, it could have also prevented some health professionals from electronically attending. Internet or e-mail attendance regarding list-servers and bulletin boards does not require formal registration for password access. Another difficulty encountered by the virtual event users was the electronic, randomly generated alphanumeric passwords, which were difficult for the user to interpret via e-mail communication (e.g., 0 or zero). The password situation was possibly further obstructed by double handling and possible inaccurate interpretation by the administrator or attendee.

USAGE

Attendance at any telecommunication "virtual event" is dependent on each member having a provider, an Internet address, and communication connection software. These requirements of themselves are still a prohibitive obstacle for many health care professionals who simply do not have access via their workplace or home computers to the telecommunications highway. The HIC'95 virtual conference event was officially attended by 80 participants, 40 authors and 40 attendees, although some 1,200 Internet attendees visited or surfed to the virtual conference site on the WWW within the 12-week period.

The conference contributing authors were invited to attend, given allocated passwords specifically to answer any questions generated from their abstracts, and to participate. Interestingly, a random investigation indicated low general usage by health informatician authors against a much higher use by virtual Internet attendees. Further investigation showed authors had a low rate of Internet address availability for either private or organization use, whereas the conference moderators and authenticated virtual attendees were frequent users of both the event and the technology.

In-depth analysis about user frequency was not performed on this inaugural event as the project aim was to develop, implement, educate, and communicate via telecommunications for practitioner connectivity and show the way, not specifically to answer a research question.

The Internet conference remained "live" for a period of 4 weeks prior to the actual event and 8 weeks following the onsite event, a total of 12 weeks. It was of considerable interest to find the most informative dialogue took place from approximately 1 week post-on-site event till closure of the virtual event. This was felt to be due, in part, to the on-site registered delegates' education and stimulation.

OUTCOMES

The virtual telecommunication event was deemed by many health professionals to be of great importance and significance and considered worthy of repeating the following year (1996). The lessons learned were necessary from both the design team's and the society's perspective. Regarding the question of utilizing information technology, the WWW, and the Internet for information connectivity among health professionals globally, now and in the future, the event response was a resounding yes from all attendees. Many attendees, while requesting modifications to the Internet dialogue format area, stated they would welcome the availability of the event the following year, 1996. Inaugural attendee participants were all sent a copy of the full proceedings on CD-ROM. The CD-ROM was available following the on-site event. CD-ROM technology offered ease of dissemination, low cost, and further demonstration of the use of another method of enabling technology. Inquiry and registration to the inaugural virtual event by participants was acknowledged by the organizing society.

IMPLICATIONS FOR NURSING

A virtual health and nursing conference, or meeting, using a WWW computer server for stored information and the Internet or Intranet as the enabling telecommunication tool allows participants to view and exchange information with many other health professionals from different sectors and departments of the health care industry, in any time frame without the sometimes inconvenience of formal face-to-face meetings. Virtual meetings can traverse time and place, primary, secondary, and tertiary sectors of health, medical, nursing, and allied health disciplines, with administrators, academics, and government health personnel all participating. Virtual meetings include the storage of event and meeting information and the exchange of dialogue, including debate, with papers, policies, procedures, and projects, plus pointers to other health resource centers and individuals who are available globally on the information superhighway.

As was demonstrated by the many virtual attendees of this event, globally distant collegial communication and friendships can occur during an event and many continue long after it ceases to be available. This project demonstrated the reality of a telecommunication event from a global perspective and the reduction of human barriers to communications. Organizational risk takers prepared to design, assist with the construction, moderate, invite, and participate for success of a virtual meeting can expect input from many sources increasing the breadth and depth of each member's knowledge along with securing more intense information and participation. Project analysis of this event showed that dialogue and discussion do take time to develop, but with adequate time frames allocated for participation, success will eventuate.

QUESTIONS FOR THE FUTURE

The exchange of information using the Internet and WWW communication tools within a dedicated forum or meeting has much to offer the nurse professional(s) practicing in urban, rural, or remote global geographies. Imagine conducting a conference with nurses around the world on a specific practice area, administration issue, policy or procedure, curriculum or education matter, or research question with the dialogue and information injection that can and will eventuate. Technological information services beg nurses to assist themselves, the profession, and consumers in the processes of care, at the same time assisting the world environment. Why do we need to continue using both formal, face-to-face meetings and hard-copy published texts using scarce resources of the organization(s) and the world? Why can't we publish and disseminate information electronically? Seek and include the valuable additional nursing knowledge external to a group? Include interested others in our dialogue? Why can't we share the "color" of all practice settings? Why can't we learn from and with each other and obtain input to our many vexing questions concerning health care delivery and quality client outcomes? Why not utilize the telecommunication tools and vehicles available to us today and at the same time reduce the human resource time, costs, and the timbers of the world? One question that keeps reappearing: Why do students from a very early age utilize telecommunication technologies for information resourcing almost every day in their lives and yet nursing continues to keep nurse practitioners and our students in the "old ways of being and doing"? In practice settings are we really assisting ourselves and our graduates?

Today's health care delivery structures and processes call upon nursing to find more efficient and effective ways of "being and doing." Telecommunication (Internet and WWW) offers the nursing profession greater assistance in search of those better ways. To facilitate dialogue and scholarship, nurses must have access to the Internet and WWW as part of their everyday practice. It is interesting to note that today some general industries employ personnel solely to surf the WWW seeking specific information that will assist the organization and its members to develop leading-edge products and marketing strategies in the 21st century. This could be a norm for nursing libraries, health organizations, nursing divisions, and care units, who could develop their own virtual telecommunication conference events, delivering and receiving information from and to nurse professionals globally, while at the same time assisting in the instruction of students and colleagues, especially those in lesser developed countries.

It is noted in both the Australian and American media there are signs of change in telecommunication policy and direction by governments and educational bodies. Telecommunication is perceived to be the basic requirement for the immediate future, a smarter way in which to work and the possibility for quality consumer outcomes. President Clinton is reported in the media as proposing a $US100 million plan to expand the Internet to every home in America in the 21st century, while in Australia a leading educational vice chancellor told a recent prestigious meeting of academics that virtual education will make universities as we know them today obsolete, and the need is the global classroom.[3,4] Professionally, nursing needs to be very proactive in these very changing times. Communication needed for today's nurse professionals needs to be available to him or her whenever required. With global collegial communication, connectivity to knowledge, skills, and inquiry is the opportunity for future professional development and growth.

At a recent Australian Health Informatics Conference, HIC'96, the keynote speaker posed to delegates the questions of modeling the technology, envisioning and mapping the course for the future, defining prevention of illness and control and treatment modalities, calling for experts to assist, and collaborating in our own future. Opportunity is with us now; rather than responding to the direction set by others, surely accountability in the prospective development and use of information technology for contemporary nursing and quality patient care is our reality now.

REFERENCES

1. S. Sparks, "Thriving on the information Superhighway: Patients and Professionals Map the Course," Proceedings of the Fourth National Health informatics Conference, 19–21 August 1996, ed. McGuiness and Leeder, Melbourne, Australia: Health Informatics Society of Australia: 1–3.
2. T. Anderson, Alberta, Canada: WWW published paper, URL, http://www.ualberta.ca/~tanderso/paper/vc/vcintro.htm
3. Reuter, "Clinton Sets Goal of Every Home on Internet," *The Age* (Saturday, 12 October 1996): A9.
4. W. Busfield, "Threat toUnis," *The Herald Sun* Friday, 11 October 1996.

9

Social Class in the United States

America's Emphasis on Welfare: Is It Children's Welfare or Corporate Welfare?

Sally Raphel

Are you aware that, according to the Children's Defense Fund (CDF) Action Council (2003):

- Each day in America, three children are killed as a result of abuse or neglect, 183 are arrested for violent crimes, 1,455 babies are born without health insurance, 2,811 high school students drop out, and 7,611 are reported abused or neglected?
- Among industrialized countries with respect to investing in and protecting children, America ranks:
  a. 12th in living standards among our poorest one fifth of children
  b. 14th in efforts to lift children out of poverty
  c. 19th in percentage of children in poverty
  d. 23rd in infant mortality
  e. Last in protecting our children against gun violence?

- The United States has the highest inmate population in the world—more than China, whose population is four times greater?
- The prison population in Texas exceeds the total undergraduate enrollment of the University of Texas system?
- The average pay of the CEOs of the 501 biggest U.S. corporations was $9.6 million in 2002? The average compensation for a CEO of $15.5 million could pay the salaries of more than 900 child-care workers for a year.

These are tough economic times when it comes to our national budget. Yet, some believe that the current budget proposals are drastically shortchanging our neediest cohort—the children and their families. The Bush administration proposes new tax cuts on top of the $1.3 trillion tax cut enacted in 2001. Bush's 2004 proposed budget includes a new round of tax cuts of $1.3 trillion (House, 2003) over the next 10 years. The provisions will give the top 1% of Americans an average of $30,000 each. And those in the bottom fifth of taxpayers will get $6.

Every 4 minutes a child is arrested for drug abuse, and every 2 hours a child or youth is killed by a firearm. Have we become desensitized by the numbers? How do we wrap our arms around numbers such as 1 in 5, or every 3 hours, when we speak of the plight of children and youth?

A new program, Child Watch Visitation Program, by the Children's Defense Fund helps put passion with numbers and action within caring communities. The ultimate goal in every area is to improve the lives of children by promoting action on local, state, and national levels. Projects are operating across the country in Los Angeles; Tampa, FL; Washington, DC; Hartford, CT; Syracuse, NY; New York, NY; Pittsburgh, PA; and Austin, TX. These programs, which can be planned and carried out in 6 weeks, are not a one-time event but part of an ongoing series to move business and political leaders to action by having them witness firsthand the current dismal world of children. The main goals are to raise awareness, create new leaders, and inspire action. It has four major components:

1. On-site visits to programs serving children
2. Briefing by public policy experts
3. Written background materials
4. Experiential activities

Each project is planned and implemented by local organizations that support the movement to "Leave No Child Behind." At the end of the program, participants leave feeling inspired by endless possibilities in ways to help. Check out the Children's Defense Fund's Web site for more details: www.childrensdefense.org/childwatch.htm.

BUSH ADMINISTRATION BUDGET PROPOSALS AND THE CHILDREN'S GAP

Four months into the current fiscal year, Congress completed action on the FY 2003 budget that is effective until September 30. To help finance increases in some priority areas, the omnibus FY 2003 budget bill includes a 0.65% across-the-board cut for almost all domestic programs. Funding for several child welfare programs will continue at current levels (minus the across-the-board cut), including Child Welfare Services, Child Abuse Prevention and Treatment Act State grants, Adoption Opportunities grants, and Adoption Incentive Payments to states.

The president's FY 2004 budget proposes the establishment of a new child welfare financing option. The proposed budget affects program areas such as: foster care, adoption assistance, promoting safe and stable families, child welfare, social services block grant (Title XX), child care, Head Start, temporary assistance for needy families, fatherhood initiatives, community- and faith-based initiatives, behavioral health, substance abuse, abandoned infants assistance, and mental health. The budget of the Substance Abuse Mental Health Services Administration is reduced from $68 million to $34 million. Medicaid/SCHIP is changed drastically: projected 2003 spending was $162 billion with funding of $4.7 billion, and projected 2004 spending is $176 billion with $5 billion budgeted. Although the proposed Medicaid-Chip block grant will provide increased funding in the initial year, funds decrease in later years and leave states with a permanent cap on federal aid. The 2004 proposal would cut 30,000 children from child-care assistance this year (200,000 over 5 years), cut 570,000 children from after-school programs, and create new bureaucratic barriers for disadvantaged children to get school lunches.

The proposed budget does include a $100 million increase for the Promoting Safe and Stable Families program, first-time funding of $42 million for educational and training vouchers for youth leaving foster care at age 18 years, and $10 million for the Mentoring of Children With Incarcerated Parents. However, funding for the Title V Local Delinquency Prevention Grant program was cut from $94.3 million to $46.5 million. This program is the only federal funding specifically targeted toward local delinquency primary prevention efforts. It funds programs that target high-risk youth who have not had contact with law enforcement.

States participating in the Adoption and Safe Families Act will be required to continue to (a) maintain the child protections outlined in the Adoption and Safe Families Act, (b) agree to maintain existing levels of state investment in child welfare programs, and (c) conduct an independent third-party evaluation of their programs. According to Secretary Thompson, "this plan would allow states to choose a fixed allocation of funds over a five-year period rather than the current entitlement funding for the Title IV-E Foster Care program" (Child Welfare League of America, 2003).

Particular Concerns

Overall concerns are that the 2004 proposal budget dismantles core services for low-income children and families under the guise of state flexibility, merges the Children's Health Insurance Program (CHIP) and Medicaid into a new block grant, and turns the Section 8 housing voucher program into a block grant that would impose a minimum new charge of $50 a month for rent, no matter how low the family's income. States would be denied the flexibility to exempt families from the new minimum charge, and would be required to get approval from Washington to do so.

Child-care services for low-income children would be frozen in place for another 5 years. After-school services for children and youth would be cut by nearly $400 million in FY 2004. The President's No Child Left Behind education bill is more than $1 billion below the level promised for 2004, and while the 2003 State of the Union Address announced $450 million for mentors for junior high school students and children whose parents are incarcerated, the 2004 budget provides only $150 million.

The 2004 budget falls $6.15 billion short of the $18.5 billion planned for Title I of the Elementary and Secondary Even Start, which provides literacy help to at-risk children and families. Individuals with Disabilities Education Act (IDEA) preschool grants for children with disabilities are frozen. The new plan also cuts $81 million from programs to improve state and local teacher quality and puts hurdles in the way of families receiving help through School Lunch and School Breakfast programs. The Elementary and Secondary School Counseling Program and the Dropout Prevention Program would be eliminated, and the grant program to help migrant students get high school diplomas or equivalency degrees is cut by more than 40%. A $50 million cut is proposed for the State Safe and Drug Free Schools program.

Advocates point out that the budget moves revenues into the pockets of the richest Americans and away from a broad range of services and supports for low- and moderate-income working families.

SUMMARY

States are facing a collective budget shortfall of $40 billion and may have to cut millions from programs that serve children and families. The 2004 proposal exacerbates the state fiscal crisis. Because of linkages between federal and state taxes, federal tax cuts have added to the loss of state revenue coming into states to assist with programs. In addition, the federal government is likely to cut human service funding to shore up its $157 billion deficit, and many nonprofit agencies are seeing donations slow dramatically because of the lagging economy and stock market.

It seems evident that the proposed federal budget cutbacks exacerbate state reductions in children's services, worsen the federal budget deficit, greatly increase the nation's debt, and mortgage America's future by passing the lack of investments in youth potential onto the next generation.

Expanded details on all of the above can be found at www.cwla.org/advocacy/2004bushbudgetchildren.htm and www.children'sdefense.org/budget_analysis.php.

REFERENCES

Child Welfare League of America. (2003). FY 2003 final budget and Bush budget 2004 [press release]. Retrieved April 7, 2003, from www.cwla.org/newsevents/news03022llu.hym

Children's Defense Fund Action Council. (2003). It's time for new voices for new choices which truly leave no child behind. Retrieved April 7, 2003, from www.cdfactioncouncil.org/2003ActionGuide.pdf

Children's Defense Fund. (2003). Bush administration wages budget war against poor children. Retrieved April 7, 2003, from www.childrensdefense.org

House Budget Committee-Democrats. (2003). Irresponsible tax agenda drives budget into permanent deficit. Retrieved June 17, 2003, from www.house.gov/budget_democrats/congressional_budgets/fy2004/conf_summary/de

10

Race and Ethnicity

The Origins and Demise of the Concept of Race

Charles Hirschman

Physical and cultural diversity have been salient features of human societies throughout history, but "race" as a scientific concept to account for human diversity is a modern phenomenon created in nineteenth-century Europe as Darwinian thought was (mis)applied to account for differences in human societies. Although modern science has discredited race as a meaningful biological concept, race has remained as an important social category because of historical patterns of interpersonal and institutional discrimination. However, the impossibility of consistent and reliable reporting of race, either as an identity or as an observed trait, means that the notion of race as a set of mutually exclusive categories is no longer tenable. As a social science term, race is being gradually abandoned. Physical differences in appearance among people remain a salient marker in everyday life, but this reality can be better framed within the concept of ethnicity.

To modern eyes, especially American ones, the reality of race is self-evident. Peoples whose ancestors originated from Africa, Asia, and Europe typically have different appearances in terms of skin color, hair texture, and other superficial features. Although racial differences may be only skin deep, it is widely assumed that races have been a primordial source of identity and intergroup antagonism from the earliest societies to the present, with ancient hatreds, exploitation, and discrimination among the most common patterns. Even in modern societies, which have exposed the myth of racism, race remains a widely used term for socially defined groups in popular discourse—and, in some countries, also in scholarly research, and public policy.

A basic problem with this perspective is that it is increasingly difficult to define and measure race as a social category. Are Jews a race? What about Muslims in Europe or Koreans in Japan? If Filipinos and Samoans are official races listed in the US census form, why can't Arab Americans or Middle Easterners be included? And how might the golfer Tiger Woods respond to the standard question about his racial identity?

Although these questions may seem merely pedantic, many critical issues of public policy are shaped by the perceptions of racial identities and racial boundaries. Who should be eligible for preferential admission to universities in the United States, Canada, Malaysia, India, South Africa, and other societies that have affirmative action policies? What are the rules for defining the descendants of indigenous peoples who are seeking redress for the expropriation of their ancestral lands in the United States, Canada, and many other countries around the globe? Who decides one's racial origins—are they based on subjective identity or are there objective criteria that observers can use? These are challenging questions that will tie policymakers and scholars into knots in the coining years as they attempt to take race into account in order to fashion nonracial or postracist societies.

In this essay, I review the history of the concept of race and its ties to social science, including demography. My conclusion (drawing on the work of other scholars) is that race and racism are not ancient or tribal beliefs but have developed apace with modernity over the last 400 years and reached their apogee in the late nineteenth and the first half of the twentieth century. Social science did not originate the belief that innate differences are associated with racial groups, but many social scientists in the Social Darwinist tradition were complicit in the construction and legitimation of racial theories.

In the twentieth century, social scientists made strident efforts to challenge the assumptions and reveal the lack of empirical evidence behind the racial theories of humankind. However, it took epochal events, most notably the specter of Nazi Germany and the nationalist movements of colonized peoples, to weaken the grip of racism as a popular and scientific theory. Although biological theories of race have been largely discredited by these political events and scientific progress, racial identities, classifications, and prejudices remain part of the fabric of many modern societies. I maintain that social science, and demography in particular, have an obligation to show that it is impossible to discuss the issue of race with any logic or consistency without an understanding of the origins and characteristics of racism.

THE ORIGINS OF PHYSICAL AND CULTURAL DIVERSITY

Modern human beings (*Homo sapiens*) are the most recent branch of hominids that emerged in Africa around 100,000 to 150,000 years ago (Cavalli-Sforza et al. 1994; Diamond 1993; Oppenheimer 2003). As humans became the dominant species in their initial ecosystem, they experienced reproductive success that increased their numbers relative to local food supplies.

The most common response to population growth in excess of the carrying capacity of a local environment is migration to new regions and ecosystems. Over the millennia, human settlements spread to most of the major regions of the world (Cavalli-Sforza et al. 1994; Davis 1974; McNeill 1984). Although archeological and genetic evidence is not entirely consistent or conclusive, the general consensus is that humans left Africa less than 100,000 years ago and reached Asia and Australia around 70,000 to 74,000 years ago, West Eurasia about 40,000 to 50,000 years ago, the Americas around 15,000 to 30,000 years ago, and finally some of the small Pacific Islands only within the last millennium (Cavalli-Sforza and Cavalli-Sforza 1995: 122; Diamond 1997: 341; Oppenheimer 2003: 348–351).

Human migration and settlement of the major world regions are not continuous processes, but are generally prompted by climate change and subsequent changes in physical geography and the availability of sustenance. Population implosions, contractions, and disappearance in local areas may have been more common than periods of expansion and dispersal. The Last Glacial Maximum, about 18,000 years ago, made many regions of the Earth uninhabitable, but the accompanying lowering of the oceans created expanded regions of human settlement in Southeast Asia and opened the Bering Straits. At the end of the last ice age, rising sea-levels drowned large areas of land, and many human communities were lost or driven to migrate to other regions (Oppenheimer 1998).

These migration waves, followed by long stretches of immobility and isolation, gave rise to human diversity. Cultural diversity arose naturally as people learned to adapt to new climatic zones and to survive on different flora and fauna. In the short term, it often appears that cultural patterns are unchanging, repeated generation after generation. Socialization is a powerful means of cultural continuity and can be remarkably effective in a stable environment. Yet human communities are capable of rapid social and cultural change in response to new environmental conditions (Diamond 1997: chapter 17). Language divergence is a natural process fostered by isolation. If two populations with a common language become separated, different dialects will arise naturally within a few centuries, and mutually unintelligible languages will develop within 1,000 to 1,500 years (Cavalli-Sforza and Cavalli-Sforza 1995: 165).

Human diversity in physical features—phenotype—also arises if populations are geographically separated from each other for long periods of time. Some external features, such as skin color and body size and shape, are highly subject to the influence of natural selection in response to climate. Areas with greater exposure to sun, such as the tropics, provided an advantage to persons with naturally darker skin pigmentation, who were more likely to have survived and to have left greater numbers of descendants in successive generations. In northern latitudes with less sunlight, cereal eaters do not receive sufficient Vitamin D, and fair skin provides a survival advantage because it allows for greater absorption of ultraviolet rays, which aids in the production of Vitamin D (Cavalli-Sforza and Cavalli-Sforza 1995: 93–94).

The distribution of different phenotypes ("races") in the modern world provides only an approximate guide to their geographical origins. Major waves of prehistoric migration have contributed to the spread of some peoples from their "place of origin." Moreover, many modern-day peoples are admixtures of different populations. For example, most populations in the Americas, north and south, reflect the very recent migrations of peoples from Europe, Africa, and Asia, as well as the blending of these peoples with native Amerindians. Similarly, the populations of modern Africa reflect major migration waves on the continent over the last 5,000 years (Diamond 1997: chapter 19).

There are different readings of the archeological, linguistic, and genetic evidence on the ancestral origins of modern-day peoples. For example, Cavalli-Sforza and Cavalli-Sforza (1995: chapter 6) argue that most contemporary Europeans are descendants of migrants from the Middle East about 10,000 years ago. Because of their early development of agriculture, peoples from the Middle East were able to expand their numbers in Europe relative to the indigenous hunting and gathering populations. However, Oppenheimer (2003: 252), relying on recent research by Richards et al. (2000), concludes that migrants from the Near East contributed only about a quarter of the genetic heritage of European populations over the last 8,000 years.

Regardless of the origins of variations in human diversity between populations, whether in phenotype or culture, there is no fixed pattern of outcomes when peoples come into contact with each other. In some cases, benign curiosity has led to peaceful accommodation, while in other cases fear, hostility, and conflict have ensued. In some instances, groups have developed ideologies of inherent superiority and inferiority. A survey of historical and contemporary societies provides a preliminary assessment of the temporal and structural antecedents of racial ideologies.

CONCEPTUAL PRELIMINARIES: THE DISTINCTION BETWEEN ETHNOCENTRISM AND RACISM

In hunting and gathering or agricultural societies, strangers are generally feared. It is necessary to overcome an initial sense of distrust between groups before it is possible to engage in symbiotic relationships of exchange, trade, or other kinds of human relationships. At a societal level, this general sense of fear, distrust, and social distance between peoples is captured by the notion of ethnocentrism. Simpson and Yinger (1985: 45) define ethnocentrism as the nearly universal tendency to believe in the rightness of one's own group and the natural aversion to difference. Ethnocentrism is a product of socialization into the beliefs and practices of one's own society, seeing them as natural and, by contrast, seeing the behavior and culture of those who are different as unnatural.

Ethnocentrism may have some basis in the natural predisposition to favor members of one's own kin group (or imagined kin group) over others. Ethnocentrism may also have "functional value" as means of reinforcing social solidarity. Recent experimental psychological research has shown that anger can create prejudice against an artificially defined alien group (DeSteno et al. 2004). In situations of conflict, it is easier to motivate persons to attack others who speak differently and who may have strange patterns of diet, beliefs, and customs. Patriotism, the celebration of a society's virtues, and the disparagement of the backwardness and the savagery of others have their roots in ethnocentrism.

Hostile and threatening behavior based on ethnocentrism is generally directed at the supposed manifestations of "otherness." The underlying logic is that other people are not like us because they have not been socialized into our language and culture. If the outsiders were to give up their foreign ways, they could (and would) become members of our society. For example, the children of the enemy are often "adopted" by conquerors after they have slaughtered the adults. The children are reared, socially and culturally, to become members of their adopted society.

Ethnocentrism, while hardly benign, is quite different from the belief that neither "others" nor the "descendants of others" could ever become like us. This alternative structure of belief, which I label "racism," holds that otherness is not simply a product of socialization, language, or culture, but is part of the inherent character of different groups. In modern terminology, racism is the belief that all humankind can be divided into a finite number of races with differing characteristics and capacities because of their genes or other inherited biological features. Therefore, adopted children inherit the attributes of their biological parents (and ancestors) and can never become the equals of their adoptive families or society.

The distinction between ethnocentrism and racism does not hinge on the presence of antipathy, the often-observed outbreaks of mass slaughter of "others," or the degree of domination and exploitation. Racism is a structure of belief that the "other community" is inherently inferior and lacks the capacity to create a society comparable to one's own. My argument is that ethnocentrism is a common feature of most societies, but that racism is a modern development of the last few centuries.

RACE AND CULTURAL DIFFERENCES IN ANCIENT AGRARIAN AND MARITIME EMPIRES

Studies of the great civilizations of antiquity, including Egypt, Greece, and Rome, show an awareness of racial features in art and literature, but little suggestion of modern forms of racism, as defined above. In summarizing the prevailing view among scholars, Snowden (1983: 63) notes:

> The ancients did accept the institution of slavery as a fact of life; they made ethnocentric judgments of other societies; they had narcissistic canons of physical beauty; the Egyptians distinguished between themselves, "the people," and outsiders; and the Greeks called foreign cultures barbarian. Yet nothing comparable to the virulent color prejudice of modern times existed in the ancient world . . . black skin was not a sign of inferiority; Greeks and Romans did not establish color as an obstacle to integration in society.

The primary activities of elites in early agrarian and maritime empires were warfare and trade. Both activities led to cross-cultural contacts, sometimes creating opportunities for alliances across ethnic divisions. In the Mediterranean world, recurrent exchanges took place between lighter-skinned peoples of lower Egypt and darker-skinned peoples of the upper Nile, which included lands known as Kush, Nubia, and Ethiopia. Persons with African features are often displayed in a positive light in paintings, statues, and other art produced by Egyptian civilization. Although wars occurred between these regions, there is evidence that Egyptian pharaohs took Nubian women as concubines and that black warriors married Egyptian women (Snowden 1983: 40–41). Later writings from Greek and Roman sources suggested a generally positive view of Africans, a respect for their way of life, and admiration for their military and political roles in the Mediterranean world (Snowden 1983: 58–59).

One of the major debates in classical studies is the influence of Egyptian civilization, and by extension African culture, on the development of Greek Hellenistic civilization. Martin Bernal (1987, 2001) asserts in *Black Athena* that the Afro-Asiatic roots of the Greek world via Egypt were subsequently minimized, if not erased, by racist European thinking of the seventeenth and eighteenth centuries. Critics of the Black Athena thesis believe that Bernal has overstated the case for Egyptian influences on the development of Greece (Lefkowitz and Rogers 1996). Bernal and his critics are, however, in essential agreement that modern racial ideology, as opposed to everyday nationalism (ethnocentrism), was largely absent from the classical world. Hannaford (1996: chapter 2) concludes not only that racial thinking was absent from the Greek understanding of humankind, but that the alternative theory of politics—the division between societies governed by ethics and morals and those ruled by barbarism—was the central tenet of classical Greek thinking.

Although most people in premodern times lived in isolated villages with relatively little contact with different cultural traditions, this was not the case with the great cities of agrarian and maritime civilizations in the Middle East, Europe, and Asia. Conquest, trade, and the creation of administrative bureaucracies produced multicultural populations of rulers, slaves, merchants, and pilgrims from distant lands, though there were different neighborhoods defined by language, religion, or region of origin. The old city of Jerusalem contains areas that are still labeled by their historical identity—Arab, Jewish, Armenian, and Christian quarters. The great Southeast Asian maritime cities of Malacca, Batavia, and Manila were divided into enclaves delineated by ethnic origins, including Chinese, Malays, Bengalis, Japanese, and Europeans. Other examples can be drawn from the great trading cities of Africa and China.

Referring to early medieval Spain, Kamen (1997: 2) describes the patterns of segregation, repression, inequality, and frequent conflicts between Christians, Jews, and Muslims, but he also notes the "existence of a multicultural framework [that] produced an extraordinary degree of mutual respect. . . . Communities lived side by side and shared many aspects of language, culture, food, and dress, consciously borrowing each other's outlook and ideas."

Although cultural misunderstandings were probably frequent in these high-density preindustrial cities and occasional acts of violence occurred between peoples of different backgrounds, strong incentives also existed to maintain order and peaceful relations. Trade and exchange, by definition, foster interdependence. Many people were drawn to cities with the hope of economic gain, primarily through trade in precious metals, cloth, spices, and other valued commodities. The most lucrative trade was often with persons who were from the most distant lands, and probably most culturally and physically dissimilar. While acts of violence, including theft and murder, might have yielded short-term economic gains, continual conflict could scarcely be the basis of a long-term commercial relationship.

Multicultural cities fared well if there were brokers who could communicate in multiple languages and understood how to work through and around cultural differences. Individuals who could fulfill the role of cultural brokers were central to the success of trading cities. The children of "mixed marriages," exposed to multiple languages and cultures during childhood, probably represented the largest share of cultural brokers. Such mixed marriages were common in all ancient cities, because traders and warriors from other lands typically married local women and produced populations of mixed descent.

The numbers of persons involved in these early cross-cultural encounters tend to be underestimated because their descendants were gradually absorbed by the host population. Some documents suggest that black persons were very common in the military campaigns throughout the Mediterranean world (Snowden 1983: 65–66). Through marriage and relationships with local women, there may have been a sizable contribution of African genes to the modern populations of Italy and other countries of southern Europe (and vice versa). There is evidence that early Chinese migrants to Southeast Asia produced descendants who were acculturated and assimilated into local populations of modern Indonesia and the Philippines (Skinner 1960, 1996; Doeppers 1998). Most of the population of contemporary Latin America and the Caribbean is an admixture of peoples of European, African, and Amerindian heritage.

In addition to population movements and trade, another important factor that created intergroup acceptance and acculturation was the spread of major religions. For many centuries, early Christianity espoused a vision of the world in which color and national origin were considered insignificant (Snowden 1983: 1081). It was the acceptance of faith that created the "City of God" or the major division of humankind, not geography, culture, or descent (Hannaford 1996: 95). The early spread of Islam and Buddhism may also have represented integrative social movements that did not place much significance on the physical and cultural differences of those who accepted the faith.

Hannaford (1996: 114–115) suggests that Christianity, and European civilization, began to change with the change in the treatment of Jews in the thirteenth century. By limiting the political tolerance for Jews in Spain, which they had enjoyed for a millennium, and forcing them to wear distinctive dress and identifying badges, Christianity began to abandon the idea of inclusiveness. The Spanish Inquisition began in 1480 with the task of attacking heresy and blasphemy, but it soon turned to rooting out Jewish converts to Christianity for not being pure Christians (Hannaford 1996:122–123; Kamen 1997: 57). The test used by the Inquisitors was "purity of blood" based upon genealogies, which were often fictive.

Abandoning the principle of a civic culture of peoples with multiple religious faiths (and even the desirability of conversion), Spain expelled hundreds of thousands of Jews in the late fifteenth century and more than one million Muslims from 1502 to 1510 (Hannaford 1996: 124–126; Kamen 1997: 23; Rehrmann 2003: 51). The methods used by the Inquisitors and their assumptions about purity of blood foreshadowed the rise of racism (Hannaford 1996: 100–104; Fredrickson 2002: 31–35). Some of the practices during the Inquisition included discriminatory harriers that precluded the children of Jewish converts to Christianity from holding official positions and professional occupations, but Kamen (1997: chapter 11) observes that these barriers were not always enforced and were sometimes contested.

THE ORIGINS OF IDEOLOGICAL RACISM

The "origins of racism" is such a broad topic that I can only highlight some key aspects of the question. My objective here is simply to make an argument that is illustrated with a few historical examples. The claim I make is that racism (defined above as the belief that social and cultural differences between groups are inherited and immutable) is a modern idea that emerged in recent centuries as a result of three transformations that created sharp divides between Europeans and other peoples: 1) the enslavement of millions of Africans in plantation economies in the New World; 2) the spread of European colonial rule across the world, especially in Asia and Africa in the nineteenth century; and 3) the development of Social Darwinism—the pseudoscientific theory of European superiority that became dominant in the nineteenth century. The word "race" and comparable terms in other languages do not appear before the late seventeenth century. It took at least another century, not until after the American and French revolutions, before the term acquired anything like its modern connotation (Hannaford 1996: 5–6).

Paralleling the fifteenth- and sixteenth-century debate about whether Jewish converts to Christianity could (and should) be assimilated as part of the Christian community in Spain was the question of how to comprehend the nature of the peoples "discovered" by Spanish explorers in the New World. Some Spanish intellectuals and officials thought that Amerindians were lesser beings who should be enslaved, while others argued, on the basis of religious principles, that all humankind shared common capacities and attributes (Hannaford 1996: 150). This debate probably did little to lessen the harshness of the conquest, exploitation, and suppression of New World peoples by Spanish conquerors. Nonetheless, the key point is that there was no universal theory of race differences that sanctioned such cruelty and mistreatment.

A variety of ideas afloat in the seventeenth and eighteenth centuries were precursors to the racial ideology of the nineteenth century. Among the most important eighteenth-century developments were the efforts by Carolus Linnaeus, Johann Friedrich Blumenbach, and Comte de Buffon to classify all flora and fauna, including humans, in a systematic framework based on morphology and complexity. In various systems, humans were classified into subspecies on the basis of geography, skin color, and physical traits (Banton 1998: chapter 2). In the pre-Darwinian era, however, there was no single authoritative account of human diversity. For example, many Europeans regarded blackness or dark skin in a negative light, but many European travelers commented positively on the physical appearance of Africans and their intelligence and abilities (Adas 1989: 66–67).

In Christian Europe, the Biblical story of creation remained the touchstone of intellectual discourse. Although some of the new theories of human diversity classified Africans and Native Americans as separate populations with lesser qualities than Europeans, there was no clear explanation of how humans had diverged over the millennia since Adam and Eve. In the centuries following the Age of Discovery that began in the late 1400s considerable debate took place over the origins of physical and cultural diversity, but no universal agreement was reached (Harris 1968).

The intellectual currents surrounding race and racial classifications were profoundly changed in the mid to late nineteenth century in the wake of Darwinian theory (Banton 1998: chapter 4). Charles Darwin presented a plausible account of the origins of species differentiation in response to environmental change. Although Darwin's theory was controversial and contested, it found immediate acceptance among intellectuals who were searching for a convincing explanation (and scientific justification) for racial differences among humans. According to this thesis, races (or geographically isolated populations) had evolved into separate subspecies over time. With the veneer of modern science and the purported evidence on variation in intelligence by cranial size, the emerging school of physical anthropology was devoted to identifying different races and their different capacities and endowments (Gould 1996). Throughout the latter half of the nineteenth century and the first half of the twentieth, Social Darwinism, eugenics, and scientific racism were among the leading ideas of science and popular culture (Harris 1968; Higham 1988; Hofstadter 1955; Fredrickson 2002).

These interpretations and theories about the origins of human diversity were not simply intellectual issues but were rooted in the reality of growing European military, economic, and political dominance (Adas 1989). Perhaps the most critical early development was the creation of plantation slavery in the New World. Slavery was a common phenomenon in many ancient societies and persisted in Asia and Africa until fairly recent times. However, there were major differences between these various forms of traditional slavery and the modern form in plantation economies.

Traditional slavery resulted from conquest and dominance of other societies and the raiding of tribal peoples who lived at the margins of agricultural and maritime empires. In Southeast Asia, slavery was closer to indentured servitude, with persons becoming slaves if they could not pay their debts to local notables (Reid 1983). Although inherently exploitative, traditional slavery was tempered in varied ways in feudal societies. For example, the children of slaves could often become assimilated into the dominant population. In traditional societies, there was often a shared history and culture between masters and slaves that did not deny the humanity of slaves. Western slavery of Africans in the New World developed as a very different institution, one that played a critical role in the emergence of the racial ideology of white supremacy. The enslavement of Africans was the defining feature of the plantation economy in the New World from the sixteenth to the nineteenth century.

Western slavery was part of an emerging world capitalist economy in which millions of Africans were transported across the Atlantic to produce sugar, tobacco, and cotton on plantations in the New World, and the products were sold to Europe. Thompson (1975:117) argued that the plantation was a race-making institution:

> The idea of race is a situational imperative; if it was not there to begin with, it tends to develop in a plantation society because it is a useful, maybe even necessary, principle of control. In Virginia, the plantation took two peoples originally differentiated as Christian and heathen, and before the century was over it had made two races.

In addition to the differences in color and culture between white masters and black slaves, the factory model of production on plantations created both spatial segregation and a rigid hierarchy. White plantation owners often delegated the harsh disciplining of slaves to foremen and other intermediaries. The extreme levels of economic exploitation combined with the denial of even the most basic human rights could only be explained by dehumanizing the enslaved population, who could be bought and sold at the whim of their owners. In the United States, this interpretation was given the legal sanction of the Supreme Court in the 1857 Dred Scott decision, which ruled that slaves were property and not entitled to the rights of citizens. Such conditions were fertile ground for the development of an ideology of white supremacy (Fredrickson 1987; Jordan 1968, 1974).

The other "race-making" institution in the modern world was European imperialism. Since the late fifteenth century, Europeans had engaged in exploitative and colonial activities in much of the world. The conquest of the Western Hemisphere, first by the Spanish and Portuguese and later by the French and the British, was a relatively easy task. Encountering less technologically advanced peoples and aided by the effect of infectious diseases on the native population, Europeans quickly mastered and largely depopulated the New World and proceeded to exploit its natural resources and develop plantation economies. The conquest of Asia and Africa was a different story because these areas were densely settled, and Europeans often encountered deadly tropical diseases that inhibited large settler populations.

Not until the nineteenth century, with the beginnings of the industrial revolution, were European powers able to dominate, militarily and politically, the landmasses and peoples of Asia and Africa. European colonialists created sharp divisions of prestige, power, and economic status between the rulers and the ruled in the Victorian Age. Because these divisions coincided with differences in color and other physical attributes between whites and the peoples of Asia and Africa, racism provided a powerful legitimation of imperialism. Adas (1989: 275, 318–319) argues that European racial ideology had relatively little impact on colonial decision making during the first half of the nineteenth century, but by the last decades of the century the assumption of the biological superiority of Europeans was deeply interwoven with debates over colonialism. Said (2003) argues that colonial hegemony created an intellectual school—Orientalism—that put scholars and academic specialists who studied colonized societies into the service of the imperial West. He contends that Orientalism advanced precisely with the expansion of direct European colonial domination from 1815 to 1914 (Said 2003:41).

This age of high imperialism—late nineteenth to mid-twentieth century—coincided with the spread of popular education, increased social mobility, and the development of democratic institutions in Europe and other regions of European settlement. The assumption of the white man's burden to provide protection and tutelage to his brown- and black-skinned brothers rationalized the discrepancies between democracy at home and authoritarian colonialism abroad. The presence of European racist beliefs in the colonies can be observed in a variety of ways, from the classifications of race and ethnicity in population censuses to the writings of colonial administrators (Hirschman 1986, 1987). These ideas, which posit that observed cultural differences reflect the heritable attributes of "races," became part of the intellectual landscape in many colonies or near colonies. Labeling the overseas Chinese as the "Jews of Asia" and rural peoples as "lazy natives" was propagated by Europeans through modern education, and ideas and vocabularies of racial distinctions were widely adopted as part of local belief structures in many parts of the world.

Although Europeans may have originated racial beliefs, these ideas became deeply rooted among colonized peoples and even in noncolonized societies. Some scholars have argued that racial consciousness in some form existed in non-Western societies before the arrival of Europeans, but there is little doubt that racial ideology came to be more deeply entrenched in the late nineteenth and early twentieth century as notions of Social Darwinism were fused with indigenous beliefs of lineage and descent (Dikotter 1992, 1997).

The racist thinking that accompanied European expansion to the colonies was built on the bedrock of earlier debates over the nature of races within Europe and the encounter with native peoples of the New World. In southern Europe, the history of the Spanish Inquisition and the expulsion of Jews and Muslims were influential, as was the struggle between Christian Europe and the Muslim Ottoman Empire. For the British (and their descendants overseas), the colonization of Ireland and the labeling of the Irish as a separate race helped to set the stage for racial ideology in other British colonies (Garner 2004; Ignatiev 1995).

Although the formal institutions of slavery were abolished in the nineteenth century in the Americas, and colonialism was beginning to be questioned and even resisted during the early decades of the twentieth century, racism had developed a life of its own. Not only did white supremacy provide economic and psychological benefits to whites, but racial ideologies were central doctrines of the modern world. Racism and the inevitability of racial inequality were affirmed by science and were widely held beliefs among most intellectuals, including leading scholars in the social sciences, well into the twentieth century. Adas (1989: 318) observes, "By the early 1900s, the eighteenth century belief in the unity of humankind found few adherents among European intellectuals and politicians." Although racist ideas may not have directly caused slavery, imperialism, and cases of modern genocide, the ideology of white superiority provided legitimation and rationalization for them.

The monstrous evils that led to the deaths of 6 million Jews in Nazi Germany and the racial apartheid system of South Africa are often considered aberrations of the twentieth century. However, racism and modernity are compatible. Germany was perhaps the most modern society in early-twentieth-century Europe and the Jewish population in Germany was largely assimilated into German culture, with a high degree of intermarriage (Fredrickson 2002: 125). Moreover, the Holocaust was carried out in modern bureaucratic fashion and included the use of modern technology. Fredrickson (2002: 104) argues that modernization is a precondition for an overtly racist regime.

Most modern states do not become racist regimes on the scale of Nazi Germany or the apartheid structure of South Africa or the pre-civil rights southern United States. But modernization does not appear to inhibit the conditions that lead to racism. For example, popular anti-Semitism in Germany was paralleled by similar beliefs in most other Western countries, including the United States. The rise of prejudices against minorities, foreigners, Catholics, and Jews was characteristic of the United States in the late nineteenth and early twentieth century (Higham 1988; Lieberson 1980). For example, in the early 1920s, Columbia, Harvard, and other Ivy League colleges enacted quotas to lower the number of Jewish students (Higham 1988: 278). There is little evidence to support the claim of an inevitable conflict between racial ideologies and modernity.

THE DECLINE OF IDEOLOGICAL RACISM

Specific historical events, in addition to the growing weight of scientific knowledge, greatly reduced the influence of official racism over the second half of the twentieth century. George Fredrickson (2002: 127), a distinguished historian of comparative race relations, concludes that the period following World War II was the turning point in the modern history of racism (also see Omi 2001). The ideology and policies of Nazi Germany crystallized the issue. As the understanding that the Holocaust was the stepchild of the doctrine of anti-Semitism began to sink into popular consciousness, a "moral revulsion" took hold that unnerved many people with "polite" racial, religious, and ethnic prejudices.

Following the defeat of fascism, formerly weak political, social, and intellectual movements that sought to change the official structures of dominance and hierarchy have gathered strength throughout the world. The most important of these was the anticolonial movement, which led to the independence of dozens of new states in Asia and Africa. The British generally cooperated in peaceful transitions of power, as did the Americans, but the French and Dutch tried to hold onto their colonies by force. Regardless of countries' initial reaction to the demands from the colonies for independence, the age of imperialism was over for all colonizing powers (save the Russian Soviet imperium) by the middle of the twentieth century. Within 15 years after World War II, almost all of the former colonial countries were independent or in the transition to independence.

When subjugated peoples gain national power or independence, especially if they had to fight for it, there is generally an insistence of equal recognition and standing in the international community. These sentiments were aided by a new international normative structure with the founding of the United

Nations and the adoption, in 1948, of the Universal Declaration of Human Rights. Part of the change was ideological, but it also coincided with a shift of real power in the world whereby European (and North American) political and economic dominance was no longer absolute.

The post-World War II rise of antiracial movements was also important within societies. The civil rights movement in the United States, led by African Americans, was the single most important force in transforming the country from its racist moorings to a more open society. The civil rights movement, which became the model for many others, was inspired by the nonviolent ideology of the anticolonial movement in India and was strongly supported by many Jewish Americans who saw the parallels between the anti-Semitism of Nazi Germany and the mistreatment of African Americans in the United States.

Science has also played an auxiliary role in the decline of racism. In the early decades of the twentieth century, a number of books written by leading academics and "public intellectuals" were sympathetic to racist views of society, and in particular deplored the deleterious influence of the massive wave of immigrants from eastern and southern Europe. This tradition is exemplified in the writings of John R. Commons (1907), an influential economist of the early twentieth century, and E. A. Ross (1914), one of the most prominent sociologists of the time. In addition, popular works professed a much more blatant expression of scientific racism, such as Madison Grant's (1916) *The Passing of the Great Race*. These voices of authority were communicated to the reading public through popular magazines and newspapers (for overviews, see Baltzell 1964; Higham 1988; Gould 1996).

There was a counterweight to racist science, most notably advanced by Franz Boas, a leading anthropologist of the early twentieth century, whose students included Margaret Mead, Melville Herskovits, and Ruth Benedict (Boas 1934: 25–36; Barkan 1992: 76–95; Stocking 1982). Before World War II, however, the liberal nonracist interpretation was just another theoretical perspective, and many scholars and most of the educated public held other views. One sign of the strength of scientific racism was the eugenics movement, which popularized the fear that the higher fertility of the lower social classes and inferior races would bring ruin to Western civilization (Haller 1984). These ideas were influential in promoting immigration restriction in the United States in the 1920s.

After World War II, science began to assert more forcefully and in a more unified voice that race and racial categories held little scientific meaning. Ashley Montagu's *Man's Most Dangerous Myth: The Fallacy of Race,* first published in 1942, became a standard text for university students in the 1950s and 1960s. UNESCO, a branch of the United Nations, issued an authoritative report in 1952 entitled *The Race Concept: Results of an Inquiry* that was intended to expose the fallacious assumptions of racial ideologies. Social science evidence on the harmful effects of racial segregation was cited in the 1954 decision of the United States Supreme Court that ended the doctrine of "separate but equal" in public schooling.

Decolonization, social movements to end discrimination, the empowerment of racial minorities, and a more antiracist social science agenda were all instrumental in moving the United States and other countries away from the apartheid-style racism that was prevalent during most of the first half of the twentieth century. Race still mattered, but the official ideology of racism was on the decline.

RACES IN A POSTRACIST WORLD

Although official racism was on the decline in the late twentieth century, prejudicial attitudes and discrimination lingered. In the United States it was not until the mid-1960s that discrimination in employment, housing, and public accommodations was made illegal. Although blatant and openly practiced discrimination is probably on the wane, many minority group members in the United States continue to encounter unfair treatment from society's gatekeepers as well as everyday slights that tear at the fabric of civility. Skin color and other attributes of physical appearance are still commonly used as identifiers for discriminatory treatment and group identity (Fredrickson 2002; Hirschman 1986; Telles 2004).

The question arises, then, of what races mean without a theory of racism to define it. Recall that before the rise of racial ideology, race or physical differences were considered to be an important dimension of human differentiation, but not qualitatively dissimilar to other social and cultural characteristics, such as ethnic origin, religion, and language. The development of "scientific" (or psuedo-scientific) racism created a new ideology that differed fundamentally from traditional ethnocentrism. The unifying idea of the new racial ideology was that racial groups were identifiable by distinct aspects of physical appearance and by innate characteristics, such as temperament, predispositions, and abilities. The preoccupation of scientific racism was to identify the number of races and their attributes. In Nazi Germany, the task focused on identifying Aryans, Jews, and other European races. In early-twentieth-century America, racial ideology was concerned with enforcement of the "one-drop rule" for persons of mixed African-European descent (whereby all persons of mixed white and black ancestry were treated as solely black), the threat posed by immigration of inferior races from Asia and Europe, and how to classify American Indians. Colonial administrators in Southeast Asia pondered the questions of whether Eurasians should be recipients of white privilege and whether the energetic Chinese race would overwhelm the indigenous natives. During the late nineteenth and early twentieth century, there were no doubts that races were real entities based on biological attributes.

As the world changed in the second half of the twentieth century, racism waned but the formerly defined races and racial boundaries remained meaningful social categories in many societies, influencing both popular perceptions and the design of public policy and scientific research. This was especially the case in the United States.

Assuming that popular attitudes can be ascertained from surveys, most white Americans no longer subscribe to biological theories of racial differences, although they do not always condone specific government efforts to lessen the effects of discrimination (Schuman et al. 1997: chapter 3; Bobo 2001). Whereas racism is on the decline, the concept of race remains central to American discourse on diversity. Racial groups are regarded as real entities and the word "race" is still widely used in the media, by academics, and by the broader public.

Although the historical assumption that populations consist of a finite number of mutually exclusive racial groups is no longer tenable, awareness of this fact is relatively new. The race and status of persons of mixed African and European ancestry was a topic of much debate in nineteenth-century America, and there were variations from state to state in how they were classified (Davis 1991). But as the one-drop rule became dogma in the late nineteenth century, all "in-between peoples" were classified as black. This became the law of the land with the 1896 US Supreme Court decision, Plessy vs. Ferguson, which reflected the notion that "belief in the natural reality of human races and racial animosities was a central tenet of the new scientific doctrine of racial differences" (Klinker and Smith 1999: 99). The United States' binary classification of black and white was not the pattern elsewhere in the Americas, where more variegated perceptions of race and class evolved (Harris 1964).

Popular perceptions of fixed racial boundaries were belied by considerable inter- and intragenerational mobility across racial boundaries. "Passing," or the surreptitious movement across racial communities, was a long-standing practice among light-skinned persons of mixed African and European descent. Recent photographs of the reunion of the descendants of Thomas Jefferson and Sally Hemmings (Jefferson's mistress) show how some of the families have become white while others chose to remain within the black community (Dao 2003). Passing from a minority to the "white" majority identity was also practiced by many Hispanics, American Indians, and even second-generation immigrant whites. The most celebrated cases are the adoption of Anglo-Saxon-sounding stage names for Hollywood stars with ethnic roots in eastern and southern Europe: Bernard Schwartz becomes Tony Curtis, Issur Danielovitch becomes Kirk Douglas, and Dino Crocetti becomes Dean Martin (Baltzell 1966: 47). Examples are also evident from the world of eminent sociologists. Meyer Schkolnick adopted the stage name Robert Merton for his early career as an amateur magician at the age of 14 (Merton 1994). William Form was originally Uli Formicola; his first name was Americanized in school and his father shortened the family name (Form 2002).

The present differs from the past in the extent of intermarriage across group boundaries and the declining urge to identify solely with the traditionally defined higher-status group. Marriage across ethnic and religious boundaries among white Americans has become normative (Alba and Golden 1986). About one-fifth to one-third of Hispanic and Asian Americans marry someone from a different racial group (Smith and Edmonston 1997: chapter 3; Farley 2002: 39; Bean and Stevens 2003: 239). Mixed marriage among African

Americans, although still at relatively low levels, has increased significantly in recent decades (Farley 1999; Kalmijn 1993; Qian 1997; Stevens and Tyler 2002). Although the tradition was to assume that children of mixed marriages would follow the "one-drop" rule in racial identification, this is no longer the case.

These mixed marriages have produced increasing numbers of Americans for whom racial identity is a matter of choice. Parents of mixed-race children exercised considerable variation in choice when reporting the "race" of their children in censuses and surveys (Waters 1999, 2002; Xie and Goyette 1997; Saenz et al. 1995). Some parents chose the race of the mother, others chose the race of the father, while others tried to report both races, even when the instructions permitted only one race to be reported on the form. These "only one race is permitted" forms (similar forms are used in schools and in other institutions) were the primary stimulus that gave rise to a social movement among mixed-race families to create a new "multiracial identity" that would allow their children to affirm their heritage from both parents. In many inquiries, nontrivial proportions of the population refuse to report a race or ethnicity or simply report that they are Americans (Lieberson and Waters 1993).

THE PROBLEM OF MEASURING RACE: THE US EXPERIENCE

The growing confusion over the meaning of race in the United States is most evident in efforts to classify the population by race in censuses or in administrative records collected by schools, hospitals, and other public agencies. In earlier times, the racial ideology of biological differences led to an obsessive concern with tracing the ancestries of persons who might straddle racial boundaries. The enumerators of the 1890 US census were instructed to collect race data as follows:

> Write white, black, mulatto, quadroon, octoroon, Chinese, Japanese, or Indian, according to the color or race of the person enumerated. Be particularly careful to distinguish between blacks, mulattoes, quadroons, and octoroons. The word "black" should be used to describe those persons who have three-fourths or more black blood; "mulatto," those who have three-eighths to five-eighths black blood; "quadroons," those persons who have one-fourth black blood; and "octoroons," those persons who have one-eighth or any trace of black blood. (US Census Bureau 2002: 27)

By 1960, the ghost of racism had almost been vanquished from the census concept of race, and users of the data were instructed about the definition of race as follows:

> The concept of race as used by the Bureau of the Census is derived from that which is commonly accepted by the general public. It does not,

therefore, reflect clear-cut definitions of biological stock, and several categories refer to national origin. . . .

Negro [includes] persons of Negro or mixed Negro and white descent . . . and persons of mixed American Indian and Negro descent unless the American Indian ancestry predominates. . . .

American Indian [includes] full blooded American Indians, persons of mixed white and Indian blood . . . if enrolled in a tribe . . . or regarded as Indians in their community. . . .

Japanese, Chinese, and Filipino, etc., are based largely on country or area of origin, and not necessarily on biological stock. (US Bureau of the Census 1963: x)

In spite of the intent to create a racial classification based on social rather than biological categories, the logic of the 1960 census measurement continued to rely upon the assumption of the one-drop rule for persons with any African ancestry. The only exception was for persons of mixed black and American Indian descent for whom American ancestry predominates—presumably this means that they "looked" American Indian in terms of skin color and other physical attributes. Was this a throwback to racist notions or simply an allowance for social identity to override the one-drop rule? Note that persons of mixed American Indian and white ancestry who are not members of tribes or recognized by members of the community were not to be classified as American Indians.

These may have been the assumptions of those who designed the census; however, Census Bureau enumerators did not enforce them. The 1960 US census was the first in which the majority of householders filled out the questionnaires themselves. The Census Bureau reported that there were only small differences in the racial composition of the population in the 1960 census that would not be consistent with projections based on the 1950 census, when the population was classified by race according to the observations of enumerators (US Bureau of the Census 1963: xi). This observation was inconsistent, however, with the increase of almost 50 percent in the number of American Indians (Snipp 2003). The color line between black and white was more clearly demarcated by the one-drop rule than for American Indians of mixed ancestry.

Following the passage of the civil rights laws in the 1960s, a new imperative emerged in measuring race in census and administrative data. Federal laws made discrimination illegal, and wide racial disparities could, in certain circumstances, be interpreted as potential evidence of discriminatory behavior. The 1965 Voting Rights Act gave the federal government the right to review electoral boundaries in areas where the potential voting power of racial groups and language minorities might be diluted by local government bodies. The 1965 act left the definitions of protected groups implicit, but 1975 legislation specified that in addition to blacks, the law was intended to protect the rights of "persons who are American Indian, Asian American, Alaska Native, or of Spanish heritage" (Edmonston and Schultze 1995: 147–148). These new federal responsibilities could only be undertaken with detailed census data on race and other groups by geographic areas (Edmonston and Schultze 1995: chapter 7; Edmonston et al. 1996:4–15).

In the wake of the civil rights revolution, the federal government, universities, corporations, and many organizations in the United States began to respond not only to the mandate to eliminate discriminatory procedures in admissions, hiring, and promotions, but also to take "affirmative action" to ensure that individuals who had encountered (or might encounter) discrimination would be included in the pool of eligible candidates to be admitted, hired, or promoted. All of these initiatives were thought to require "objective" data on race and ethnicity gathered by census-style questions in surveys or administrative records.

By the 1970s, the census measurement of race was still in the process of transition from race as a biological concept to a social category, based on the subjective identification of respondents who were supposed to check the right categories or to fill in the blanks according to the everyday understanding of race. Awareness was also growing, however, that race was a "political" category in the sense that the numbers of a group mattered. In addition to the formal considerations of legal or judicial mandates, benefits are associated with a population whose presence can be counted. The leadership of minority communities recognized the significance of being counted and began to push for inclusion in the census. In the 1970 census, a new question on Hispanic origin was included on the long form (5 percent sample), and in 1980 Hispanic origin was moved to the short form (100 percent of the respondents). This was a complete turnaround from the response to the creation of a category for Mexicans in the race question in the 1930 census. In the 1930s, the Mexican American population (and the Mexican government) protested against the effort to stigmatize Mexican Americans by labeling them a racial group (Cortes 1980). By the 1970s, the leadership of the Latino community understood that census data on their population was more of an asset than a liability, although they insisted on being labeled an ethnic group rather than a racial one.

A similar political effort was made in the late 1980s to ensure a detailed listing of specific Asian and Pacific Islander populations on the census form. The Census Bureau proposed a global Asian and Pacific Islander category, with a blank line so that individuals could write in their particular national origin. Representatives of the Asian American community argued that the Bureau's proposal might lead to a lower count of their populations, and, with the help of their Congressional representatives, they were successful in expanding the list to include eight specific national-origin populations (and an "other" category) on the 1990 census form (US Bureau of the Census 1990).

With a growing awareness that there was no clear conceptual framework within which to collect data on race and ethnicity in the United States, in 1977 the US Office of Management and Budget (OMB) stepped in and issued

Statistical Directive No. 15, "Race and ethnic standards for federal statistics and administrative reporting" (Edmonston et al. 1996: Appendix B). Statistical Directive No. 15 is long on details about the specific groups to be included and the format of data presentation, but it provides minimal information on the concepts of race and ethnicity or on the logic of why some groups and not others are included in the classifications. The introduction notes that "these classifications should not be interpreted as being scientific or anthropological in nature, nor should they be viewed as determinants of eligibility for participation in any Federal program." The only rationale offered for the classification is that it was developed in response to needs of the executive branch and Congress. In defining the five major race/ethnic populations to be measured (American Indian or Alaska Native, Asian or Pacific Islander, Black, Hispanic, and White), Statistical Directive No. 15 refers to persons having origins in specific regions of the world. Persons with mixed ancestry or origins should be classified in "the category which most closely reflects the individual's recognition in his community."

The listing of categories offered by Statistical Directive No. 15 was ad hoc, there were no clear criteria to define race or ethnicity, and there were no directions on how people were to be classified (presumably according to subjective identity, but this was not specified). Perhaps most importantly, political considerations were clearly beginning to tear at any consensus that the OMB classification was the best possible one. Pacific Islanders, especially Hawaiian natives, felt that their inclusion with Asians was inappropriate. Many ethnic communities or national origin populations, such as Arab Americans, thought that they should be included as populations designated on the census form. Most vocal was the multiracial population, who thought that the mutually exclusive categories of the race classification forced persons of mixed ancestry to choose only one (Farley 2002).

To address these and many other concerns, OMB, in the early 1990s, requested that the National Research Council convene a workshop to receive input from researchers, administrators, and other interested parties on a revision of Statistical Directive No. 15 (Edmonston et al. 1996). There were also studies by the Census Bureau, opportunities for public comments, and extensive interagency discussions that led to the final revision of Statistical Directive No. 15, issued in 1997 (Office of Management and Budget 1997a, 1997b). The results were ad hoc modification of the race and ethnic categories and their labels (American Indian or Alaska Native, Asian, Black or African American, Hispanic or Latino, Native Hawaiian or Other Pacific Islander, and White) and a major change in measurement that allowed persons to "mark one or more" races with which they identify (for a comprehensive overview of multiple-race classification, see Perlmann and Waters 2002). There was no additional conceptual clarification of the meaning or definition beyond the 1977 statement that race and ethnicity do not represent scientific or anthropological categories.

One might draw rather different conclusions about the significance of these varied approaches to measuring race as a social category. On the positive side, one might say that most Americans are able to think in terms of, and to reliably report themselves in, discrete race and ethnic categories. The overwhelming majority of African Americans, Asian Americans, and whites will classify themselves in the same categories regardless of what classifications are used (Hirschman, Alba, and Farley 2000). While the measurement of race changed from the basis of enumerator observation in the 1950 census to householder choice in the 1960 census, there were few signs of this change as reflected in consistency of reporting. In the 2000 census with the major change to allow for multiple race identification, only 2.4 percent of persons chose to mark two or more race groups (Grieco and Cassidy 2001). Even without conceptual clarity of the meaning of race and despite wide variations in measurement strategies and the possibility that the tabulation of the population by race will add to more than 100 percent, it seems that race is so deeply embedded in the American psyche that the society can continue to think about and measure race as it has always done.

However, an alternative case can be made that the conceptual validity of race is beginning to wane. In the 1990 census, about 5 million Americans (2 percent of the population) did not answer the race question, and another 10 million wrote in a response that did not appear in one of the listed categories (Edmonston et al. 1996: 21–22). Coding these "write in" responses was a major task with about 300 codes for race, 600 codes for American Indian tribes, and more than 70 for Hispanic groups in the 1990 census. In 1990 more than 250,000 census respondents wrote in roughly 75 different multiple race responses under the "other race" category (Edmonston et al. 1996: 22). The Census Bureau recoded many of these write-in responses to one of the listed categories. Most of those left in the "other race" category were primarily of Hispanic origin. Most census users are probably unaware of the resulting "untidiness" of responses to the traditional single "forced-choice" race question.

There are two populations for whom the race question is particularly problematic: American Indians and Hispanics. There is a broad continuum of persons of American Indian descent, whose identity varies with the nature of the question asked. For example, Matthew Snipp (1989: chapter 2) has shown that only about one in five persons who claimed American Indian ancestry (which allows for multiple responses) in the 1980 census reported themselves to be American Indian in the census race question (which had mutually exclusive categories). Mixed ancestry, which may represent the future of other racial populations, has been a characteristic of the American Indian population for at least a century, perhaps even longer (Snipp 2002). Mixed marriage rates among the American Indian population are higher than for any other racial group (Sandefur and McKinnel 1986). This means that the count of American Indians from any census and survey depends on subjective identity and the specific question asked.

With the lessening of the stigma (and the increase of the advantages) of being American Indian, a virtual population explosion in their numbers has occurred that cannot be explained by natural increase and immigration (Eschbach 1993, 1995; Passel 1996). The 1990 American Indian census population of 1.96 million grew to 2.46 million in 2000 if only the single identifiers are included and to 4.12 million if multiple identifiers are included (Snipp 2002; Grieco and Cassidy 2001). The ten-year increase may be either 24 percent or 110 percent! An increasing proportion of the mixed ancestry population has switched their "race"—signifying that being American Indian is now viewed positively (Snipp 1997).

Although explicitly not labeled as a racial group in Census Bureau classifications, Hispanic or Latino origin has a quasi-racial status by virtue of its mention in Statistical Directive No. 15 and in the 1997 revision. Because Hispanic origin is defined by a separate question, it is possible to cross-classify Hispanics by race and to compare white Hispanics, black Hispanics, Asian Hispanics, and so on. In most analyses, however, Hispanics are lumped together regardless of their race, perhaps because many Hispanics do not consider the American racial categories to be meaningful. Of the 35 million Hispanics enumerated in the 2000 census, 42 percent checked "some other race" in response to the race question, often adding an Hispanic or Latino descriptive term in the write-in category (Greico and Cassidy 2001: 10). Another 6 percent of Hispanics checked multiple races, with the second race being "some other race," indicating that they considered Hispanic origin as equivalent to the checked race category.

Another sign of the lack of recognition of American racial categories among Hispanics is the fraction that simply does not answer the race question. In the 1995 Race and Ethnic Targeted Test (RAETT) survey, about 13–14 percent of persons in the targeted Hispanic sample (living in areas that were predominately Hispanic) left the race question blank (Hirschman, Alba, and Farley 2000: 338–339). However, nonresponse to the race question dropped to less than one percent when Hispanic was included as a category in a combined race/ethnic classification. Whether because many recent Hispanic immigrants are not socialized into thinking in terms of American racial groups or because many Hispanics feel that their ancestry/language/ethnic group is their only identity, the conclusion is that members of America's largest minority community do not agree with the traditional means of designating race (Rodriguez 2000). Mainland Puerto Ricans perceive no association between skin tone (reported by interviewers) and subjective "race"—being either white or black (Landale and Oropesa 2002).

Although the Census Bureau is entrusted with measuring continuity of the country's statistics, the Bureau has an erratic record in the measurement of race. The 1980 census dropped the word "race" from the questionnaire and simply asked respondents to complete the sentence "This person is—" by choosing from a list of racial categories. The word "race" was brought back in the 1990 and 2000 censuses, presumably because respondents needed to know how this question differed from the other census items that measure Hispanic origins and ancestry. Another fundamental and growing problem is the inherent subjectivity of race responses. Statistics are supposed to be about objective phenomena that can be measured with some reliability. Under the present American system, however, a person's race is whatever the respondent thinks she or he is—in response to a list of categories. If the list of races, the names of races, or the sequence of questions changes, people's responses will change as well. These problems were exacerbated with the freedom of multiple-race reporting in the 2000 census. The preliminary report of multiple-race reporting from this census lists 57 possible combinations of 2, 3, 4, 5, or 6 races.

CONCLUSIONS

The thesis of this essay is that the common view—officially sanctioned in some countries, notably the United States—of race as a fundamental and relatively stable ascriptive attribute of human populations is flawed. To put the matter simply, there is no conceptual basis for race except racism. Racism (or racial ideology) assumes that aspects of physical appearance—phenotype—are outward manifestations of heritable traits such as abilities, propensities for certain behaviors, and other sociocultural characteristics. This assumption, though widely accepted only a couple of generations ago, has been put to rest by recent genetic research.

> It is because these differences are external that these racial differences strike us so forcibly, and we automatically assume that differences of similar magnitude exist below the surface, in the rest of our makeup. This is simply not so; the remainder of our genetic makeup hardly differs at all. (Cavalli-Sforza and Cavalli-Sforza 1995: 124)

Although overt racism has receded (though prejudice and discrimination have not been eradicated) and racial ideologies have lost official and scientific legitimacy, race remains part of the popular and scholarly lexicon, especially in the United States. The term is now viewed as a social—not a biological—category to describe members of a population who share some common physical features (e.g., skin color) and whose ancestors share a common geographical origin.

This new concept of race as a social category, however, does not have a logical basis. In general the definition tends to rely on popular perceptions of race—race is whatever people think they are or whatever they think others are. This approach, however incoherent or inconsistent, usually works in societies that have only recently emerged from under the shadow of official racism.

For example, the contemporary distinction in the United States between black and white rests on the assumption of the one-drop rule that all persons of mixed white and black ancestry are considered and treated as solely black.

This arbitrary classification reflects popular perceptions, but only because most Americans, white and black, have been accustomed to think this way. Most other societies with mixed white-black populations, such as in Latin America or South Africa, do not have a one-drop rule, but rather consider color to be a continuum or have other groupings. There is no correct way to classify people of mixed ancestry within the standard American race categories—popular perceptions vary across societies and over time, reflecting historical experiences of official racism and state-sanctioned discrimination. With intergroup relations in flux, including high levels of intermarriage that are producing even larger populations of mixed ancestry, popular perceptions of race will surely change, although we have only vague perceptions of what the future might hold.

On this very shaky ground American society has created social arrangements and public policies that assume that race is a real phenomenon and that distinct racial populations exist. These arrangements and policies are a reflection of public and private efforts to combat the legacy of the era of official segregation and discrimination. During that era, casual observation of physical appearance was sufficient for gatekeepers to deny housing, employment, and promotion opportunities to those considered "nonwhite."

The contemporary situation is very different. Racial and gender discrimination is illegal, and administrative bodies and courts are expected to evaluate the evidence when institutional discrimination is alleged. One standard method of ascertaining discrimination is to examine statistical proportionality—the difference in racial composition in an economic sector or residential area from that which might be expected if discrimination were absent (Prewitt 2002). Given the new method of collecting data on multiple race identities in the 2000 census, several experts believe that it will become much more difficult to enforce civil rights laws and to use the method of statistical proportionality as evidence of institutional discrimination that will hold up in courts (Harrison 2002; Persily 2002).

A related problem is evident when gatekeepers (human resource directors, college admission committees, supervisors) make affirmative action efforts to ensure some measure of opportunity for those who may have (or whose ancestors have) encountered discrimination. Who is eligible for such consideration? How are such efforts to be weighed against other attributes, including the potential for achievement and other personal circumstances that may indicate hardship? How will others, such as courts of law, the news media, and public opinion, perceive decisions made by gatekeepers? In these circumstances, many idiosyncratic factors will affect individual and institutional behavior. Organizations that wish to report progress in minority hiring may take the broadest definition of racial groups and include recent immigrants whose ancestors were not subjected to official discrimination in American society. On the other hand, gatekeepers may try to be risk-averse and inquire about an applicant's ancestry to ensure that he or she is a legitimate candidate for such programs. Because these programs are controversial and opportunities and resources (though modest by any measure) are being allocated, some clear basis must exist for making decisions and evaluating them.

Perhaps the most telling example of the difficulty of reconciling social programs that rely on unambiguous definitions of race and the demographic reality of mixed ancestry is the situation of American Indians. As a token effort to compensate American Indians for the expropriation of their lands and periodic programs of physical and cultural extermination, the Bureau of Indian Affairs, a branch of the Department of Interior, provides subsidies for health care and other programs. Individual American Indians are eligible only if they are recognized as members of tribes and the US government recognizes their tribe. Since most criteria for tribal affiliation and the recognition of a tribe are inherently subjective, most tribes (and the government) have fallen back on genealogy, expressed in terms of the infamous "blood quantum" rules that are a legacy of nineteenth-century racist thought (Snipp 2002). Because American Indian tribes have a degree of sovereignty, it has been possible to establish profitable gambling casinos on some Indian reservations that are close to major metropolitan areas. This new industry has raised the economic value of American Indian affiliation considerably. The result is a growing number of new "Indian wars" (political struggles) between existing and nascent tribes over the question of recognition and potential wealth, with the debate often focused on "blood quantum" as the measure of true racial membership.

There has never been any credible justification for assuming that physical markers, such as skin color, can be considered as ascriptive characteristics that universally predict sociocultural characteristics. It was only during the era of scientific racism of the century or more before 1950 when laws and social policies oppressed nonwhites that racial divisions could be considered as relatively permanent divisions of the population. Contemporary societies, including the United States, now reject the racial ideologies of the past, but US practice continues to try to measure race as an ascribed status and to fashion social policies as if racial categories are mutually exclusive. The 2000 US census that allowed multiple race reporting has exposed the fault line in the system, but the full implications of what has transpired have yet to be acknowledged. Most players appear to favor a technical fix to the problem—an improved system of measurement that would be comparable to the past and better reflect the present. The reality, in my judgment, is that the concept is broken and there is no valid rationale for preserving the old system, however modified. Race without racism is an anachronism.

There is a perfectly good concept to take the place of race, that of ethnicity. Ethnicity is explicitly subjective, it acknowledges multiple ancestries, and it recognizes that ethnic groups are porous and heterogeneous. Physical differences in appearance among people remain an important marker in everyday life, but this reality can be better framed within the concept of ethnicity, which emphasizes ambiguity rather than either/or distinctions. Although jettisoning the concept of race is a necessary step, this act alone will not solve the problem and may make it more difficult to measure inequality and discrimination in the short run (e.g., Hirschman 1993; Hirschman, Alba, and Farley 2000:391). The more important and challenging task is to develop meaningful measures of ethnicity and alternative concepts for use in censuses, surveys, and administrative records.

REFERENCES

Adas, Michael. 1989. *Machines as the Measure of Men: Science, Technology, and Ideologies of Western Dominance.* Ithaca: Cornell University Press.

Alba, Richard D. and R. M. Golden. 1986. "Patterns of ethnic marriage in the United States," *Social Forces* 65:202–223.

Baltzell, E. Digby. 1964. *The Protestant Establishment: Aristocracy and Caste in America.* New York: Vintage Books.

Banton, Michael. 1998. *Racial Theories.* Second edition. Cambridge: Cambridge University Press.

Barkan, Elazar. 1992. *The Retreat of Scientific Racism.* Cambridge: Cambridge University Press.

Barth, Fredrik (ed.). 1969. *Ethnic Groups and Boundaries.* Boston: Little, Brown.

Bean, Frank and Gillian Stevens. 2003. *America's Newcomers and the Dynamics of Diversity.* New York: Russell Sage Foundation.

Bernal, Martin. 1987. *Black Athena: The Afroasiatic Roots of Classical Civilization, Volume 1. The Fabrication of Ancient Greece, 1785–1985.* London: Free Association Books.

———. 2001. Edited by David Chioni Moore. *Black Athena Writes Back: Martin Bernal Responds to his Critics.* Durham: Duke University Press.

Boas, Franz. 1934. "Race," in Edwin R. A. Seligman (ed.), *Encyclopedia of the Social Sciences, Vol. 13.* New York: Macmillan, pp. 25–36.

Bobo. Lawrence D. 2001. "Racial attitudes and relations at the close of the twentieth century," in Neil J. Smelser, William Julius Wilson, and Faith Mitchell (eds.), *America Becoming: Racial Trends and Their Consequences. Vol. 1.* Washington, DC: National Academy Press, pp. 264–301.

Cavalli-Sforza, L. Luca, Paolo Menozzi, and Alberto Piazza. 1994. *The History and Geography of Human Genes.* Abridged paperback edition. Princeton: Princeton University Press.

Cavalli-Sforza, Luigi Luca, and Francesco Cavalli-Sforza. 1995. *The Great Human Diasporas: The History of Diversity and Evolution.* Cambridge, MA: Perseus Books.

Commons, John R. 1907. *Races and Immigrants in America.* New York: Macmillan.

Cortes, Carlos E. 1980. "Mexicans," in Stephan Thernstrom (ed.), *The Harvard Encyclopedia of American Ethnic Groups*. Cambridge, MA: Harvard University Press, pp. 697–719.

Dao, James. 2003. "A family get-together of historic proportions," New York Times, July 13, p. 1.

Davis, F. James. 1991. *Who Is Black? One Nation's Definition*. University Park: Pennsylvania State University Press.

Davis, Kingsley. 1974. "The migrations of human populations," in *The Human Population. A Scientific American Book*. San Francisco: W.H. Freeman, pp. 53–65.

DeSteno, David, Nilanjana Dasgupta, Monica Y. Bartlett, and Aida Cajdric. 2004. "Prejudice from thin air: The effect of emotion on automatic intergroup attitudes," *Psychological Science* 15:319–324.

Diamond, Jared. 1993. *The Third Chimpanzee*. New York: Harper Perennial Library.

———. 1997. *Guns, Germs, and Steel: The Fates of Human Societies*. New York: W.W. Norton.

Dikotter, Frank. 1992. *The Discourse of Race in Modern China*. Stanford: Stanford University Press.

——— (ed.). 1997. *The Construction of Racial Identity in China and Japan: Historical and Contemporary Perspectives*. Honolulu: University of Hawai'i Press.

Doeppers, Daniel. 1998. "Evidence from the grave: The changing social composition of the population of metropolitan Manila and Molo, Iliolo, during the latter nineteenth century," in Daniel Doeppers and Peter Xenos (eds.), *Population and History: The Demographic Origins of the Modern Philippines*. Monograph No. 16. Madison: Center for Southeast Asian Studies, University of Wisconsin, pp. 265–277.

Edmonston, Barry and Charles Schultze (eds.). 1995. *Modernizing the U.S. Census*. Washington, DC: National Academy Press.

Edmonston, Barry, Joshua Goldstein, and Juanita T. Lott (eds.). 1996. *Spotlight on Heterogeneity: The Federal Standards for Racial and Ethnic Classification, Summary of a Workshop*, Washington, DC: National Academy Press.

Eschbach, Karl. 1993. "Changing identification among American Indians and Alaskan natives," *Demography* 30:635–652.

———. 1995. "The enduring and vanishing American Indian," *Ethnic and Racial Studies* 18:89–108.

Farley, Reynolds. 1991. "The new census question about ancestry: What did it tell us?," *Demography* 28:411–430.

———. 1999. "Racial issues: Recent trends in residential patterns and intermarriage," in Neil J. Smelser and Jeffrey C. Alexander (eds.), *Diversity and Its Discontents: Cultural Conflict and Common Ground in Contemporary American Society*. Princeton: Princeton University Press, pp. 85–128.

———. 2002. "Racial identities in 2000: The response to the multiple race response option," in Joel Perlmann and Mary C. Waters (eds.), *The New Race Question: How the Census Counts Multiracial Individuals.* New York: Russell Sage Foundation, pp. 33–61.

Form, William. 2002. *Work and Academic Politics: A Journeyman's Story.* New Brunswick, NJ: Transaction Publishers.

Fredrickson, George M. 1981. *White Supremacy: A Comparative Study in American and South African History.* New York: Oxford University Press.

———. 1987. *The Black Image in the White Mind: The Debate on Afro American Character and Destiny, 1817–1914.* Middletown, CT: Wesleyan University Press.

———. 2002. *Racism: A Short History.* Princeton: Princeton University Press.

Garner, Steve. 2004. *Racism and the Irish Experience.* London: Pluto Press.

Gould, Stephen Jay. 1996. *The Mismeasure of Man.* Revised and expanded edition. New York: W.W. Norton.

Grant, Madison. 1916. *The Passing of the Great Race: Or the Racial History of European History.* New York: Scribner's.

Grieco, Elizabeth M. and Rachel C. Cassidy. 2001. *Overview of Race and Hispanic Origin: Census 2000 Brief C2KBR/01-1.* Washington, DC: US Census Bureau.

Haller, Mark. 1963. *Eugenics: Hereditarian Attitudes in American Thought.* New Brunswick: Rutgers University Press.

Hannaford, Ivan. 1996. *Race: The History of an Idea in the West.* Washington, DC: Woodrow Wilson Center Press.

Harris, David. 2002. "Who is multiracial? Assessing the complexity of lived race," American *Sociological Review* 67:614–627.

Harris, Marvin, 1964. *Patterns of Race in the Americas.* New York: W.W. Norton.

———. 1968. "Race," in David L. Sills (ed.), *International Encyclopedia of the - Social Sciences. Vol. 13.* New York: Macmillan and Free Press, pp. 263–268.

Harrison, Roderick. 2002. "Inadequacies of multiple response race data in the federal statistical system," in Joel Perlmann and Mary C. Waters (eds.), *The New Race Question: How the Census Counts Multiracial Individuals.* New York: Russell Sage Foundation, pp. 137–160.

Higham, John. 1988. *Strangers in the Land: Patterns of American Nativism, 1860–1925.* 2nd. ed. New Brunswick: Rutgers University Press.

——— (ed.). 1997. *Civil Rights and Social Wrongs: Black-White Relations Since World War II.* University Park: Pennsylvania State University Press.

Hirschman, Charles. 1983. "America's melting pot reconsidered," *Annual Review of Sociology* 9:397–423.

———. 1986. "The making of race in colonial Malaya: Political economy and racial ideology," *Sociological Forum* 1:330–361.

———. 1987. "The meaning and measurement of ethnicity in Malaysia: An analysis of Census classifications," *Journal of Asian Studies* 46:555–582.

———. 1993. "How to measure ethnicity: An immodest proposal," in Challenges of Measuring an Ethnic World: Science, Politics, and Reality, Proceedings of the Joint Canada-United States Conference on the Measurement of Ethnicity, sponsored by Statistics Canada and the US Bureau of the Census. Washington, DC: US Government Printing Office, pp. 547–560.

Hirschman, Charles, Richard Alba, and Reynolds Farley. 2000. "The meaning and measurement of race in the U.S. Census: Glimpses into the future," *Demography* 37:381–393.

Hofstader, Richard. 1955. *Social Darwinism in American Thought*. New York: G. Braziller.

Hughes, Everett and Helen M. Hughes. 1952. *When Peoples Meet: Race and Ethnic Frontiers*. Glencoe, Illinois: Free Press.

Ignatiev, Noel. 1995. *How the Irish Became White*. New York: Routledge.

Jordan, Winthrop D. 1968. *White Over Black: American Attitudes Toward the Negro, 1550–1812*. Chapel Hill: University of North Carolina Press.

———. 1974. *The White Man's Burden: Historical Origins of Racism in the United States*. New York: Oxford University Press.

Kalmijn, Matthijs. 1993. "Trends in Black/White intermarriage," *Social Forces* 72:119–146.

Kamen, Henry. 1997. *The Spanish Inquisition: A Historical Perspective*. New Haven: Yale University Press.

Klinker, Philip A. and Rogers M. Smith. 1999. *The Unsteady March: The Rise and Decline of Racial Equality in America*. Chicago: University of Chicago Press.

Lefkowitz, Mary R. and Guy Mclean Rogers. 1996. *Black Athena Revisited*. Chapel Hill: University of North Carolina Press.

Landale, Nancy and R. S. Oropesa. 2002. "White, Black, or Puerto Rican? Racial self identification among mainland and island Puerto Ricans," *Social Forces* 81:231–254.

Lieberson, Stanley. 1980. *A Piece of the Pie: Blacks and White Immigrants Since 1880*. Berkeley: University of California Press.

Lieberson, Stanley and Mary Waters. 1993. "The ethnic responses of Whites: What causes their instability, simplification, and inconsistency?," *Social Forces* 72:421–450.

McNeill, William H. 1984. "Human migration in historical perspective," *Population and Development Review* 10:1–18.

Merton, Robert K. 1994. "A life of learning," The Charles Homer Haskins Lecture Series. Occasional Paper No. 25. New York: American Council of Learned Societies.

Montagu, Ashley. 1942. *Man's Most Dangerous Myth: The Fallacy of Race.* New York: Columbia University Press.

Myrdal, Gunnar. 1964. *An American Dilemma.* Two volumes. New York: McGraw-Hill (original publication in 1944).

Office of Management and Budget. 1997a. "Recommendations from the Interagency Committee for the Review of the Race and Ethnic Standards to the Office of Management and Budget concerning changes to the standards for the classification of federal data on race and ethnicity," Federal Register. Vol 62, No. 131, Wednesday, July 9, 1997/Notices. pp. 36874–36946.

———. 1997b. Revisions to the standards for the classification of federal data on race and ethnicity, Federal Register. Vol. 62, No. 210, Thursday, October 30, 1997/Notices. pp. 58782–58790.

Omi, Michael A. 2001. The changing meaning of race, in Neil J. Smelser, William Julius Wilson, and Faith Mitchell (eds.), *America Becoming: Racial Trends and Their Consequences.* Vol. 1. Washington, DC: National Academy Press, pp. 243-263.

Oppenheimer, Stephen. 1998. *Eden in the East: The Drowned Continent of Southeast Asia.* London: Weidenfeld and Nicolson.

———. 2003. *The Real Eve: Modern Man's Journey Out of Africa.* New York: Carroll and Graff Publishers.

Passel, Jeffrey. 1996. The growing American Indian population, 1960–1990: Beyond demography, in Gary D. Sandefur, Ronald Rindfuss, and Barney Cohen (eds.), *Changing Numbers, Changing Needs.* Washington, DC: National Academy Press.

Perlmann, Joel and Mary C. Waters. 2002. *The New Race Question: How the Census Counts Multiracial Individuals.* New York: Russell Sage Foundation.

Persily, Nathaniel. 2002. "The legal implications of a multiracial census," in Joel Perlmann and Mary C. Waters (eds.), *The New Race Question: How the Census Counts Multiracial Individuals.* New York: Russell Sage Foundation, pp. 161–186.

Prewitt, Kenneth. 2002. Race in the 2000 Census: A turning point, in Joel Perlmann and Mary C. Waters (eds.), *The New Race Question: How the Census Counts Multiracial Individuals.* New York: Russell Sage Foundation, pp. 354–361.

Qian, Zhenchao. 1997. Breaking the racial barriers: Variations in interracial marriage between 1980 and 1990, *Demography* 34:263–276.

Rehrmann, Nobert. 2003. A legendary place of encounter: The convivencia of Moors, Jews, and Christians in medieval Spain," in Dirk Hoerder (ed.), *The Historical Practice of Diversity: Transcultural Interactions from the Early Modern Mediterranean to the Postcolonial World.* New York: Berghahn Books, pp. 35–53.

Reid, Anthony. 1983. "Introduction: Slavery and bondage in Southeast Asian history," in Anthony Reid (ed.), *Slavery, Bondage, and Dependency in Southeast Asia*. St Lucia, Australia: University of Queensland Press, pp. 1–43.

Richards, Martin et al. 2000. "Tracing European founder lineages in the Near Eastern mtDNA pool," *American Journal of Human Genetics* 67: 1251–1276.

Rodriguez, Clara E. 2000. *Changing Race: Latinos, the Census, and the History of Ethnicity in the United States.* New York: New York University Press.

Ross, Edward A. 1914. *The Old World in the New: The Significance of Past and Present Immigration to the American People.* New York: Century.

Said, Edward. 2003 [1979]. *Orientalism. 25th Anniversary Edition.* New York: Vintage Books.

Sandefur, Gary and Trudy McKinnel. 1986. "American Indian intermarriage," *Social Science Research* 15:347–371.

Saenz, Rogelio, Sean-Shong Hwang, Bengino Aguirre, and Robert Anderson. 1995, "Persistence and change in Asian identity among children of intermarried couples," *Sociological Perspectives* 38:175–194.

Schuman, Howard, Charlotte Steeh, Lawrence Bobo, and Maria Krysan. 1997. *Racial Attitudes in America: Trends and Interpretations.* Revised Edition. Cambridge, MA: Harvard University Press.

Simpson, George Eaton and J. Milton Yinger. 1985. *Racial and Cultural Minorities: An Analysis of Prejudice and Discrimination.* Fifth Edition. New York: Plenum Press.

Skinner, G. William. 1960. Change and persistence in Chinese culture overseas: A comparison of Thailand and Java, *Journal of the South Seas Society* 16 (1/2):86–100.

———. 1996. Creolized Chinese societies in Southeast Asia, in Anthony Reid (ed.), *Sojourners and Settlers: Histories of Southeast Asia and the Chinese.* Asian Studies Association of Australia Southeast Asia Publications Series No. 28. St. Leonards, NSW: Allen and Unwin, pp. 51–93.

Smith, James P. and Barry Edmonston. 1997. *The New Americans: Economic, Demographic and Fiscal Effects of Immigration.* Washington, DC: National Academy Press.

Snipp, C. Matthew. 1989. *American Indians: The First of this Land.* New York: Russell Sage Foundation.

———. 1997. Some observations about racial boundaries and the experiences of American Indians," *Racial and Ethnic Studies* 20:667–689.

———. 2002. "American Indians: Clues to the future of other racial groups," in Joel Perlmann and Mary C. Waters (eds.), *The New Race Question: How the Census Counts Multiracial Individuals.* New York: Russell Sage Foundation, pp. 189–214.

———. 2003. "Racial measurement in the American census: Past practices and implications for the future," *Annual Review of Sociology* 29:563–585.

Snowden, Frank M., Jr. 1983. *Before Color Prejudice: The Ancient View of Blacks*. Cambridge, MA: Harvard University Press.

Stevens, Gillian and Michael K. Tyler. 2002. "Ethnic and racial intermarriage in the Untied States: Old and new regimes," in Nancy A. Denton and Stewart E. Tolnay (eds.), *American Diversity: A Demographic Challenge for the Twenty-First Century*. Albany: State University of New York Press, pp. 221–242.

Stocking, George W. 1982. *Race, Culture, and Evolution: Essays in the History of Anthropology*. Chicago: University of Chicago Press.

Sung, Betty Lee. 1990. Chinese American intermarriage, *Journal of Comparative Family Studies* 21:337–352.

Telles, Edward E. 2004. *Race in Another America: The Significance of Skin Color in Brazil*. Princeton: Princeton University Press.

Thompson, Edgar T. 1975. *Plantation Societies, Race Relations, and the South: The Regimentation of Populations*. Durham: Duke University Press.

UNESCO (United Nations Educational, Scientific, and Cultural Organization). 1952. *The Concept of Race: Results of an Inquiry*. Paris: UNESCO.

US Bureau of the Census. 1963. United States Census of Population 1960: Subject Reports. Non-white Population by Race, PC (2)-1C. Washington, DC: Bureau of the Census.

———. 1990. Census of Population and Housing (1990), Content Determination Reports, 1990 CDR-6: Race and Ethnic Origin. Washington, DC: US Government Printing Office.

US Census Bureau. 2002. Measuring America: The Decennial Censuses from 1790 to 2000. Washington, DC: US Government Printing Office

Waters, Mary. 1990. *Ethnic Options: Choosing Identities in America*. Berkeley: University of California Press.

———. 1999. *Black Identities: West Indian Immigrant Dreams and American Realities*. New York: Russell Sage Foundation.

———. 2002. The social construction of race and ethnicity: Some examples from demography, in Nancy Denton and Stewart Tolnay (eds.), *American Diversity: A Demographic Challenge for the Twenty-First Century*. Albany: State University of New York Press, pp. 25–49.

Xie, Yu and Kimberly Goyette. 1997. "The racial identification of biracial children with one Asian parent: Evidence from the 1990 Census," *Social Forces* 76:547–570.

11

Sex and Gender

Women, Disability, and Sport and Physical Fitness Activity: The Intersection of Gender and Disability Dynamics

Elaine M. Blinde and Sarah G. McCallister

The experiences of women with physical disabilities in relation to sports and physical fitness participation are explored. Results are discussed in terms of type of participation, reasons for participation, nature of outcomes, and gender-related aspects of participation.

Our understanding of the gender dynamics surrounding women's entry and experiences in the realm of sport and physical fitness activity has significantly expanded in recent years (Birrell & Cole, 1994; Nilges, 1997). In particular, much has been written regarding the masculine hegemony embedded in sport and the concurrent negative evaluation of women's physical capabilities in this context (Bryson, 1987; Willis, 1994). A focal point of much of this work is the gendered female body (Cole, 1994; Hall, 1996; Theberge, 1985). Perceptions of the female body as weak, inferior, and lacking physical competency are central in this evaluation process (Hall, 1996; Messner, 1988).

Research Quarterly for Exercise and Sport, Sept 1999, v70, i3, p303.

© 1999 American Alliance for Health, Physical Education, Recreation and Dance (AAHPERD). Reprinted with permission.

As a result of these perceptions and beliefs, women have encountered various forms of resistance, marginalization, trivialization, and stigmatization in their attempts to construct meaningful experiences in sport and physical fitness activity (Bryson, 1987; Theberge, 1985). These social processes reinforce the importance of assumed physical differences between men and women and promote images of male domination and superiority (Willis, 1994).

Although researchers have increasingly focused on women in the sport domain, our understanding of the sport and physical fitness activity experiences of women with disabilities is limited (Sherrill, 1993a). In particular, the intersection of gender, disability, and sport and physical fitness activity has rarely been examined (DePauw, 1997a). Gender and sport dynamics may be more complex when examining the experiences of women with physical disabilities (Blinde & Taub, 1999; DePauw, 1994). As the female body is central in determining women's marginalized role in sport and physical fitness activity, societal views of the body may be even more problematic for women with physical disabilities (DePauw, 1997a, 1997b). Perceptions of the disabled body as weak, passive, and dependent further distance it from the sport and physical fitness activity domain, where physicality is central (DePauw, 1997b).

Women with disabilities generally experience the "double handicap" of being a woman and having a disability (Deegan & Brooks, 1985). The intersection of sexism and disability discrimination situates women with disabilities in a unique social constellation (Fine & Asch, 1985). Although we have seen an increased focus on the dynamics of gender and disability (Deegan & Brooks, 1985; Fine & Asch, 1988; Hanna & Rogovsky, 1993; Lonsdale, 1990; Morris, 1993), our understanding of how these multiple forms of oppression come together in sport and physical fitness activity is limited.

Existing work exploring the experiences of women with disabilities in sport and physical fitness activity has often focused on participants in organized sport or elite-level athletes (Brasile, 1988; Brasile, Kleiber, & Harnisch, 1991; Hopper, 1986; Horvat, French, & Henschen, 1986; Sherrill, 1993b). Although research has highlighted the experiences of a small number of women with disabilities, findings cannot be assumed to generalize to the vast majority of women with disabilities who participate in less structured sport and physical fitness activities.

Henderson and Bedini (1995) conducted one of the few studies exploring the physical activity experiences of adult women with physical disabilities who were not athletes or elite-level sport participants. In their interviews of 16 women with mobility impairments, Henderson and Bedini examined how this group experienced physical activity, recreation, and leisure. In general, these women reported various perspectives from which to view the value of these activities, including leisure, therapeutic gains, and maintaining mental and physical health. On the other hand, some respondents indicated that these activities had limited value in their lives. In a related study, Henderson, Bedini, and Hecht (1994) interviewed 30 adult women with various sensory and physical disabilities about their physical activity participation. Interestingly, most of these women did not experience their bodies in active or physical ways when engaging in these activities. Rather than seeing physical activity as leisure, it was more commonly viewed as a form of therapy.

The purpose of the present paper was to enhance our limited knowledge of the experiences of women with physical disabilities in sport and physical fitness activity. Topics examined included the type and nature of participation, reasons for participation, outcomes of participation, and gendered aspects of participation. Interviews examined the intersection of gender, disability, and participation in sport and physical fitness activity.

METHOD

Participants

Participants included 16 women with a variety of physical disabilities. This sample ranged in age from 1954 years, with a mean age of 31.5 years. Fifteen of the participants were Caucasian, and 1 was African American. All the women had completed high school, 7 had earned a bachelor of science degree, and 1 had completed a master's degree. At the time of the interview, 7 of the women were enrolled in classes at either the undergraduate or graduate level. The remaining 9 participants lived in communities in southern Illinois.

The range of physical disabilities included cerebral palsy (n = 6), paraplegia (n = 4), hemiplegia (n = 2), partial paralysis (n = 1), muscular dystrophy (n = 1), spinal muscular atrophy (n = 1), and osteogenesis imperfecta (n = 1). Of these disabilities, nine were congenital, and seven were acquired after birth (with an average of 20.4 years since each had acquired the disability). Relative to these individuals' movement capabilities, 14 women regularly used a wheelchair for mobility. However, 4 of these 14 women were ambulatory if provided the assistance of a walker, crutches, or another person. The remaining two respondents could walk but were limited in terms of strength and stamina.

These 16 women had originally volunteered to participate in a grant project providing individualized recreational activities for persons with physical and sensory disabilities.(1) The grant project recruited individuals with disabilities to participate in a variety of individualized recreational programs designed to impact several aspects of the participants' lives. Participants for this program were recruited through the campus disability support service office and area community independent living centers. Fliers and consent forms were distributed by these organizations. Participants in the program were volunteers with no rewards or credit given for participation. No prerequisites existed in terms of prior participation in sport or physical fitness activity, and participants were able to select recreational activities in which they wanted to participate. Women with physical disabilities who volunteered to participate for approximately the first 16 months of this program were included in the sample (1994–95). Prior to participating in any recreational activity provided by the grant program, the 16 women were interviewed to learn about their past experiences in sport and physical fitness activity. These interviews formed the data base for the present study.

Instrument

Grant personnel developed an interview guide consisting of questions exploring the sport and physical fitness activity experiences of the participants prior to participating in the recreational program provided by the grant. Pilot interviews were conducted to assist with constructing the final interview guide. As the data collection process extended over approximately 16 months, some modifications were made in the content of the interview guide over this period (Locke, 1989).

Focus areas of the interview and sample questions within each included the following: (a) life history of the disability, (b) past participation experiences in sport and physical fitness activities, both before and after the disability, (c) reasons for participation, (d) importance and enjoyment of participation, (e) encouragement and discouragement received for participation, (f) self-perceptions of the body and one's physical capabilities, and (g) type of positive and negative outcomes of participation.

In addition to these focus areas, questions and probes were designed to facilitate an understanding of the role of gender in the participation experience. A few gender-related items were added during the ongoing interview period. Respondents were encouraged to use a broad interpretation of sport and physical fitness activity throughout the interview, ranging from organized and competitive activities to informal or causal involvement?

Procedures

All women who returned an informed consent form were contacted by telephone to arrange an interview date and time. Interviews were conducted on a university campus by two female doctoral students trained in interviewing techniques. Questions on the interview guide were open ended, and probing techniques were used to encourage respondents to discuss the most relevant aspects of their participation experiences. Interviews averaged 63 min in length and were tape recorded to provide an accurate record of the conversation. To protect the identity of participants, code names and numbers were assigned to each interviewee. Approval of the university's Human Subjects Committee was obtained before any contact with participants.

Once the interviews were completed, the tape recordings were transcribed and checked for accuracy. The two investigators independently read the finalized transcriptions. Using the process of qualitative categorization (Strauss & Corbin, 1990), the researchers attempted to identify patterns and themes in the women's responses. Each investigator clumped responses into broad categories and then organized them into subcategories (Highlen & Finley, 1996). This categorization scheme was used for each of the main issues explored in the interview. On completing this process, the two researchers compared the categories and subcategories constructed for each general section of the interview schedule. As a result of this comparison, final coding summaries were developed which contained both common and divergent patterns and included actual quotes from the women supporting or refuting a theme or pattern.

Results

On examining the interview responses, the investigators identified several interesting patterns and themes including the following general areas related to sport and physical fitness activity participation: (1) type of participation, (2) reasons for participation, (3) nature of outcomes from participation, and (4) gender-related aspects of participation (see Table 1). Actual comments from the women participants are used to highlight their experiences in sport and physical fitness activity.

Type of Participation

The women were first asked to describe the type of sport and physical fitness activities in which they had participated since their disability. All 16 discussed some form of participation, naming a variety of activities. The three most commonly mentioned activities reported included lifting weights (n = 10), swimming (n = 9), and working on fitness machines, for example, stationery bicycles (n = 8). Other activities identified were walking (n = 4), wheelchair sports (n = 4), aerobics (n = 3), bowling (n = 3), canoeing (n = 2), and fishing (n = 2). Although the women discussed many activities, the overwhelming majority of activities were fitness-related as opposed to sport-related. On the average, the women participated in these activities approximately twice a week.

Reasons for Participation

When asked why they participated in sport and physical fitness activities, the dominant theme identified in the participants' responses was maintaining body functionality. Although health and fitness concerns certainly underlie maintaining a functioning body, nearly two-thirds of the respondents specifically discussed this aspect of participation by referring to the body's ability to function. They typically expressed concerns about maintaining rather than gaining physical function. For example, Carol, a 31-year-old graduate student with cerebral palsy, stated that she wanted to "maintain what function that I do have." Similarly, when asked why she participated in sport and physical fitness activities, Sandy, a 34-year-old woman with cerebral palsy, stated, "It's, um, a more interesting way to help maintain my functional skills." Sandy went on to comment:

> As I get older I'm learning how important just plain old movement is. . . . It doesn't even have to be real structured, just, just movement in general gets circulation going and works muscles and gives other muscles a break.

TABLE 1. Outcomes of participation in sport and physical fitness activity

General outcomes	Specific patterns
1. Type of participation	Participation in fitness-related as opposed to sport-related activities
2. Reasons for participation	Maintaining functionality Social factors Psychological factors
3. Nature of outcomes from participation	Intrinsic gains: Enhanced view of capabilities Body as source of strength Motivational outlet Sense of control in life
4. Gender-related aspects of participation	Contrasting reasons for participation Contrasting societal views regarding participation

The women saw sport and physical fitness activities as a means to help preserve and maintain a body already at risk. As Kelly, a 19-year-old high school graduate with cerebral palsy, commented, "I've always had to, um, exercise, because if I don't I get weaker. My body doesn't, doesn't hold up to walk." The realization that further loss of body function might seriously alter their current standard of living was evident. Tami, a 32-year-old with paraplegia for the past 15 years, reflected on the importance of participating in sport and physical fitness activity:

> As I grow older, I want my body in better shape, because even with a disability, um, I mean, I've got that strike against me, and, uh, I just want to make sure that my body is able to carry me on through.

Alice, a 54-year-old college graduate with muscular dystrophy, also discussed what most influences her decisions to participate in sport and physical fitness activity:

> I'm really more interested, to be honest, in things that can help me as an individual at my age. I'm past the point of just wanting it for recreation. . . . Right now I'd just like to be able to maintain and keep whatever I've got left to keep life as normal as possible.

Although maintaining body functioning was cited most frequently as a key reason for participation, two other areas were noted in the women's responses. Approximately one-third mentioned social factors as an important reason for participation. In general, the sport and physical fitness activity provided these women an opportunity "to socialize" and "meet people." Connie, a 49-year-old high school graduate with cerebral palsy, high-lighted the importance of social factors in the following response when she was asked about the reasons behind her participation in activities such as lifting weights, walking, riding an exercise bicycle, and swimming:

> I just like being with people. . . . It gets me out of the apartment, you know, when you live in an apartment you can't just hibernate, you know, you've got to get out.

A few women also discussed psychological factors motivating their participation in sport and physical fitness activities. Among the psychological reasons were to "release stress" and "feel good about myself." Teresa, a 20-year-old college freshman with cerebral palsy, provided the following comment when asked why she participated in sport and physical fitness activity:

> It's like a drug; it's addictive, I think. I'm, instead of an alcoholic, I'm a sportaholic. . . . I love it, it's quite a mental high.

Teresa added that despite her love for sport and physical fitness activity, she was presently not participating, because she did not have anyone to help her with participation.

Nature of Outcomes from Participation

The various questions asked during the interview assisted the investigators in learning more about the nature of outcomes experienced from the participants' sport and physical fitness activity participation. Questions focused on areas such as positive and negative aspects of participation, meanings and feelings associated with participation, and things learned about oneself and one's physical capabilities from participation.

The one unifying theme among their responses was the intrinsic nature of what they gained from participation. Rather than focusing on extrinsic factors such as winning, status, awards, recognition, or visibility, the women almost exclusively discussed outcomes reflecting intrinsic gains of participation. Although their initial motives for participation often focused on maintaining a functional level of the body, they saw sport and physical fitness activity participation as an enlightening experience impacting several aspects of their lives. Common areas discussed during the interviews included an enhanced view of one's capabilities, seeing the body as a source of strength, viewing sport and physical fitness activity as a motivational outlet, and experiencing a greater sense of control in their lives. Despite efforts to identify both positive and negative outcomes derived from the participation experience, most responses were positive. Where appropriate, however, examples of negative experiences are noted.

Enhanced Views of Capabilities

When discussing participation outcomes from sport and physical fitness activities, one of the most commonly mentioned gains was an enhanced view of one's capabilities. Despite the physical limitations or constraints a disability can impose on an individual's lifestyle, approximately half of the women discussed how participating in sport and physical fitness activities often provided them with an expanded sense of their physical capabilities. For example, comments such as, "I can do a lot more than, you know, before [participation]," and, "I think it makes me feel more confident about what I can do," were not unusual. Similarly, the statement, "I think it helps me at times not feel like I'm quite as disabled," typified these feelings. Sandy, a 34-year-old with cerebral palsy who had received a college degree in special education, provided some interesting insight into how she felt about herself when she participated in sport and physical fitness activities:

> I tend to feel a whole lot more capable when I'm in the midst of doing something in terms of fitness and sports or, um, even if I'm not particularly successful at it, just the fact that I'm actively doing something. . . . It seems to be a real, like, freeing experience type thing for me.

Or, as Jackie, a 20-year-old college junior with paraplegia commented, "I can do more than, you know, than just pushing around [my wheelchair]."

However, perceptions of enhanced capability were not limited to sport or physical fitness activity. Again, approximately half the respondents discussed the impact of participation on their capabilities in general life situations. Examples of areas impacted included social life, community integration, work place, interpersonal relationships, and student role. These women felt that sport and physical fitness activity provided them with a renewed (or, in some cases, first-time) sense of confidence in a variety of situations. For example, when asked what she had learned about herself from participating in sport and physical fitness activity, Kathy, a 25-year-old graduate student with paraplegia for the past 10 years, stated:

> I'm probably a stronger person physically and mentally than I thought I was, that I'm more persistent than, maybe, I thought I could be.

When asked if this information will help her in other areas of life, Kathy continued:

> Oh, yeah. I think, um, in terms of, you know, just school, I think it improves my drive and my desire to succeed there. . . . I think it'll help me in the work place, urn, in dealing with other people, um, and being assertive, and also, um, in terms of commitment to work.

Body as Source of Strength
A second theme was seeing the body as a source of strength. Approximately half the women mentioned outcomes that challenged the stereotypical portrayal of disability as weakness. Their comments frequently highlighted how sport and physical fitness activity enhanced views of the body and its potential. For example, they sometimes mentioned participation outcomes such as "building up your strength" and "keeping your body fit." Jennifer, a 21-year-old high

school graduate with cerebral palsy, talked about how walking, swimming, and lifting weights made her feel:

> Good. I think it makes my body feel better, it makes . . . you know, I think from the outside and the inside, I mean, mentally and physically I think it's better.

Sandy, who previously discussed how sport and physical fitness activity enhanced perceptions of her capabilities, also commented on how she felt about her body when participating in these activities:

> It tends to be a[n] invigorating experience, and I tend to be, um, feel more positive about my body and, um, less self conscious about how I look, because I'm usually focused on specific goals.

Although many of the women talked about seeing the body as a source of strength, this perception did not characterize all the respondents' views. To the contrary, some still struggled with the assumed contradictions between disability and sport and physical fitness activity. Julie mentioned feeling "frustrated once in a while when I try and do something" and "not being able to do what I used to do." When asked how she viewed her body in this context, Wanda, a 47-year-old woman with incomplete hemiplegia for the past 26 years, responded:

> Like I view [a] calculator that's misfunctioning, well, something without all the gears, something, you know, it's not doing what it's supposed to. . . . I catch myself saying to myself, "Can we stop now?" and it's funny, because I'm the one who started it and yet I'm the one that wants to stop.

Participation as a Motivational Outlet

As respondents discussed their varied experiences in sport and physical fitness activity, several women mentioned how this context provided them with an important motivational outlet. They sometimes saw participation in sport and physical fitness activity as providing a reason "to get out of the apartment" and be "with other people." In a similar light, Cheryl, a 36-year-old with hemiplegia for 19 years, credited participation with giving her "something to look forward to when I wake-up in the morning." Connie, a 49-year-old woman with cerebral palsy, provided several related comments about the importance of participation:

> It gives me something to look forward to. . . . I just like to do it, and I think it's good for me. And I just enjoy it. . . . [It] give[s] me more to think about, and you talk about with other people.

Other women mentioned the newness or unique aspect of this form of activity. Sport and physical fitness represented variety in many of their lives. For example, Sandy talked about how participation made her feel:

> It's that feeling of, like, new challenges and new horizons and . . . just, you know, newness, whereas like physical therapy and that kind of thing. I mean, I'm 34, and I've been doing physical therapy since I was, like,

5 years old. And physical therapy, they tend not to vary the routine a whole lot, you know, whereas, I mean, recreation, I mean, it has a little different twist because you're working and you're doing physical activity.

Or, as Julie stated about the positive aspects of participation, "getting to do activities which, you know you wouldn't be doing otherwise."

Greater Sense of Control in Lives

Many women viewed participation in sport and physical fitness activity as enhancing the perceived degree of control in their lives. Their responses highlighted several different aspects of control, including attaining a sense of freedom, accomplishing what they set out to, doing something for themselves, feeling empowered, believing that no obstacles can stop them, and experiencing a. sense of mastery.

Kelly, a 19-year-old with cerebral palsy, felt strongly about what she had learned about herself through participating in sport and physical fitness activity. She stated, "I can do anything I want to. Nothing can stop me." Similarly, when asked about her overall experiences in sport and physical fitness activity, Julie, a 32-year-old college student with partial paralysis, commented:

> I feel fit. . . . I feel like I have control over it, over myself, you know, I feel, it's important. . . . It gives you a better mental health. . . . feeling happier about yourself.

Julie added that having "control over things" was important to her and that sport and physical fitness activity provided a "sense of empowerment." Wanda, a 47-year-old with incomplete hemiplegia, mentioned a "feeling of mastery" when asked what participation meant to her. Finally, Alice, a 54-year-old with muscular dystrophy who had taught school for about 14 years, commented that participation "give[s] me a certain, um, freedom."

Gender-Related Aspects of Participation

Several gender-related issues emerged in the women's responses. Questions about reasons for participation, societal expectations for participation, and views of the body often included inquiries about their perceived role of gender in the participation process. A prominent theme was their varying perceptions as to why men and women with disabilities participate in sport and physical fitness activity. As discussed earlier, the women interviewed often emphasized maintenance and health factors underlying their own participation in sport and physical fitness activity. Several respondents, however, felt that sport and physical fitness activity was a means for men with disabilities to accentuate their masculinity and enhance their male identity. Helen, a 26-year-old college freshman with osteogenesis imperfecti, highlighted the differing reasons men and women with disabilities participate in sport and physical fitness activity:

> Men are always having to prove themselves to other men, and to other guys. And I don't think girls, I mean, I don't know . . . when it comes to physical fitness, I don't think that girls are doing it for that reason.

Or, as Evelyn, a 27-year-old college junior with incomplete paraplegia, stated, "they [men] want to keep their macho attitude up."

Some women also discussed how men with disabilities were more attracted to the competitive and aggressive domain of sport than women with disabilities. For example, Barbara, a 31-year-old college graduate who was born with spinal muscular atrophy, commented on her perceptions of why men and women with physical disabilities participate in sport and physical fitness activity:

> Men are more sports oriented in general than women. . . . They're more into, like, competitive stuff, I think, than women. . . . women, it's more for health some, a lot of times, and men, they like to be, um, competitive more, like guys are competitive.

Although the women in the sample discussed the importance of developing their bodies through sport and physical fitness activity, their primary focus was on maintaining their bodies. Few comments indicated that appearance or public perception were major factors influencing participation. For example, when the women mentioned the desire to lose weight through participation, such statements were generally in the context of enhancing mobility or reducing the stress excess weight placed on their bodies. Conversely, respondents often felt that men with disabilities were more concerned with the public presentation of their bodies. These women felt that men were focused on "keep[ing] their bodies in shape" or desired "to improve their bodies."

Respondents were asked to comment on how their position as women influenced their participation. Many women discussed contrasting societal perceptions of men and women with disabilities in sport and physical fitness activity. In general, these women felt that society typically does not expect to find women with disabilities in this domain. As a result, the atmosphere surrounding their participation may lack support and be influenced by a wide range of negative perceptions and stereotypes. Julie, a 32-year-old college sophomore with partial paralysis, discussed how societal perceptions impacted her participation in sport and physical fitness activity:

> I think being a disabled female is the double whammy. . . . People are coming around a lot more to disabled males, you know, kind of in physical activity, and they're still weird to us females, you know, disabled females.

Similarly, Sandy, a 34-year-old college graduate with cerebral palsy, commented about the uneasiness she sometimes felt when working out in a weightlifting room among men and other individuals without disabilities. Her feelings of self consciousness were due in part to the emphasis placed on body building in the weightlifting room and her not looking "quite as good in terms of dress because of braces or orthopedic shoes." Sandy also used the phrase "double whammy" when discussing the complexities of being a woman and having a disability.

Although most of the women recognized the dual influence of being female and having a disability, a few respondents minimized the importance of gender in both their sport and physical fitness activity experiences and life in general. Compared to gender, the disability was seen as much more limiting in life situations. This perspective was best characterized by Carol, a 31-year-old woman with cerebral palsy who was working on her second master's degree:

> I've been socialized with the invisibility factor placed on me by society because of my disability. I do not see strong gender separation issues when it comes to people with disabilities.

DISCUSSION

The dual forces of sexism and disability discrimination obviously impact the experiences of women with physical disabilities. Many writers have highlighted how these forces create a "double jeopardy" for women with disabilities (Benefield & Head, 1984; Deegan & Brooks, 1985; Hanna & Rogovsky, 1993; Wendell, 1989). In such accounts, the everyday experiences of women with disabilities are viewed as more problematic than those of both women without disabilities and men with disabilities. The entry of women with physical disabilities into sport and physical fitness activity creates additional dynamics, as the forces of sexism and disability discrimination intersect with the societal construction of sport and physical fitness activity. Several themes emerged in the women's responses which highlight how their sport and physical fitness activity experiences may differ from those of men with disabilities and able-bodied women.

The social construction of sport and physical fitness activity is a critical factor impacting expectations and perceptions related to the participation of women with physical disabilities in this context. Sport and physical fitness activity have traditionally been constructed to accentuate virtues such as strength, aggression, independence, and physical competency (DePauw, 1997b; Whitson, 1990). This portrayal of sport and physical fitness activity often is perceived to be in opposition to the capabilities of individuals with disabilities, where assumptions of weakness, passivity, dependency, and physical inability exist (DePauw, 1997b). Despite the conflicting perceptions between sport and physical fitness activity and the context of physical disability, the dynamics of this contradiction may differentially impact men and women with physical disabilities.

Respondents perceived the gap between sport and physical fitness activity and disability as less for men with physical disabilities than for women with physical disabilities. Not only did women with disabilities experience a contradiction between the values of sport and physical fitness activity and perceptions of disability, but there was also a conflict between the female role and the construction of this activity domain. As reflected in the comments of many of women interviewed, the dynamics of a disability and being a woman distanced them from the sport and physical fitness activity context to the point that society did not generally expect to find women with disabilities in this domain. Thus, opportunities were often limited and encouragement or support lacking. As suggested by DePauw (1997b), the marginalization experienced by women with disabilities is due to their inability to match the ideals of both physicality and masculinity that are assumed to be cornerstones of sport and physical fitness activity.

In addition, many of the women interviewed generally felt that reasons for sport and physical fitness activity participation differed for men and women with disabilities. Although assumptions about a physical disability often run counter to the construction of sport, the male gender role is still perceived to be congruent with the sport domain. Several respondents thought that men with disabilities used sport and physical fitness activity as a means to prove their masculinity and reaffirm their male identity. This perception tended to correspond with actual accounts of men with physical disabilities (Taub, Blinde, & Greer, in press). Connections between masculinity and the body often are highlighted, as sport teaches men how to use their bodies in forceful and physical ways (Whitson, 1990).

Socially constructed notions about sport and physical fitness activity also differentially impact women with physical disabilities and able-bodied women. Although this construction of sport and physical fitness activity often has resulted in barriers for women in general, the degree of negativism associated with the body may not be as intense for able-bodied women compared to those with physical disabilities. The body is central, as it represents the mechanism through which physicality is demonstrated. Despite perceptions that the female body does not possess the same qualifies and potential as the male body (Hall, 1996), able-bodied women have been provided greater opportunities than women with disabilities to experience sport and physical fitness activity. For example, able-bodied women have used sport and physical fitness activity to enhance strength and fitness levels (Balsamo, 1994), develop strong and competent bodies (Blinde, Taub, & Han, 1993; Guthrie, 1997), and enhance the body's attractiveness and sexuality (Markula, 1993; Theberge, 1985).

Although such themes sometimes emerged in the responses of the women in this study, they participated in sport and physical fitness activity primarily to preserve or maintain a body already at risk. Rather than participating in sport and physical fitness activity to build a body with new capabilities and potential, many of the women acknowledged that their involvement was motivated by a desire to prevent further loss of bodily function. This finding parallels that of Henderson et al. (1994) in their interviews of adult women with disabilities. Also, comments focusing on the desire to construct a more attractive or aesthetically pleasing body were not common.

While maintenance and functionality were major motives for participation, the sense of empowerment experienced through sport and physical fitness activity was noteworthy for these women. Sport and physical fitness activity often represented a means through which women with disabilities reevaluated their capabilities and potential. Although sport and physical fitness activity participation stood in sharp contrast to societal constructions of a disability, several women talked about how participation was a "freeing experience" for them. Moreover, participation often challenged feelings of being disabled and resulted in altered perceptions of the body as a source of strength. Participation also enhanced perceptions of having "control" over certain aspects of their lives.

Sport and physical fitness activity also provided women with disabilities an intrinsically rewarding experience. Despite the range of participation outcomes these women discussed, most of the gains were foundational in nature. Being able to engage in some form of sport or physical fitness activity was often sufficient to enhance their sense of physical capabilities. The women interviewed generally did not hold high or unrealistic expectations in terms of what they desired to obtain from sport or physical fitness activity participation. In many cases, success in the activity was secondary to the fact that they were actually able to do something physical. Or, in other cases, merely getting "out of the apartment," "being with people," or having "something to look forward to" made sport and physical fitness activity extremely rewarding. Discussing outcomes at a foundational level is consistent with past research exploring the experiences of individuals with disabilities in the sport and physical fitness activity domain (Blinde & McClung, 1997; Blinde & Taub, 1999; Patrick & Bignall, 1984).

As most of the existing research on women with disabilities has focused on the experiences of individuals participating in organized sport or elite level programs (e.g., Paralympics), it is interesting to note how some themes in the present study may contrast with the voices portrayed in existing research. The women in this study participated in loosely organized and nonelite fitness-related activities and commonly discussed functionality and maintenance, as well as the foundational nature of gains derived from participation. Although elements of functionality and maintenance certainly underlie the orientations of elite level athletes, these themes are manifested much differently for the high-performance athlete. For example, comments from the women in our sample may contrast with the views expressed by female athletes with disabilities who use sport and physical fitness activities to affirm physical competence and focus attention on a person's abilities rather than disabilities (Sherrill, 1993a). Also, the responses of the 16 women in this study did not highlight the importance on factors such as competition, success, and extrinsic rewards in organized sport.

Women with physical disabilities experience the intersection of two socially devalued roles (i.e., having a disability and being a woman) in a context (i.e., sport) that by definition is constructed to run counter to both of these roles. Regardless of the specific type of disability, the body is central in the dynamics that result from the coming together of these three constructs—gender, disability, and sport and physical fitness activity. The intersection of these constructs is essential in enhancing our understanding of the sport and physical fitness activity experiences of women with physical disabilities.

In addition to considering the intersection of gender, disability, and sport and physical fitness activity, future research needs to explore in greater depth possible differences within the population of women with disabilities. For example, patterns may vary by type of disability, congenital versus acquired disabilities, or time living with a disability. Moreover, subgroups within the population of women with physical disabilities (e.g., race, ethnicity, age, social class, sexual orientation, educational background) could introduce additional belief systems that further complicate the dynamics in the sport and physical fitness activity context. Future research should be sensitive to the impact of multiple minority group status on individuals' experiences (Deegan, 1985). Taken as whole, these factors should assist in furthering our understanding of a group of women that has rarely been the focus of our research.

NOTES

1. The present study was drawn from a grant funded by the U.S. Department of Education. Principal investigators of the project were from Southern Illinois University at Carbondale—Kathleen Plesko (Disability Support Services), Diane Taub (Sociology), and Elaine Blinde (Physical Education).

2. The following statement was read to respondents before the interview to aid their understanding of the phrase "sport and physical fitness activity:"

   Sport and physical fitness activities range from competitive and organized events (e.g., wheelchair basketball, road racing) to informal and unstructured physical fitness contexts (e.g., exercise, weight lifting, swimming). These activities can include team and individual sports as well as sport and physical fitness activities that are done informally with friends or alone. Please exclude nonsport or nonphysical fitness activities having a movement component that are part of your daily routine (e.g., getting dressed, going to classes, eating, driving, working on a computer).

REFERENCES

Balsamo, A. (1994). Feminist bodybuilding. In S. Birrell & C. L. Cole (Eds.), *Women, sport, and culture* (pp. 341–352). Champaign, IL: Human Kinetics.

Benefield, L., & Head, D. W. (1984). Discrimination and disabled women. *Journal of Humanistic Education and Development, 23*(2), 60–68.

Birrell, S., & Cole, C. L. (Eds.) (1994). *Women, sport, and culture.* Champaign, IL: Human Kinetics.

Blinde, E. M., & McClung, L. R. (1997). Enhancing the physical and social self through recreational activity: Accounts of individuals with physical disabilities. *Adapted Physical Activity Quarterly, 14,* 327–344.

Blinde, E. M., & Taub, D. E. (1999). Personal empowerment through sport and physical fitness activity: Perspectives from male college students with physical and sensory sensory disabilities. *Journal of Sport Behavior, 22,* 181–202.

Blinde, E. M., Taub, D. E., & Han. L. (1993). Sport participation and women's personal empowerment: Experiences of the college athlete. *Journal of Sport and Social Issues, 17,* 47–60.

Brasile, F. M. (1988). Psychological factors that influence participation in wheelchair basketball. *Palaestra, 4*(3), 16-19, 25–27.

Brasile, F. M., Kleiber, D. A., & Harnisch, D. (1991). Analysis of participation incentives among athletes with and without disabilities. *Therapeutic Recreation Journal, 25,* 18–33.

Bryson, L. (1987). Sport and the maintenance of masculine hegemony. *Women's Studies International Forum, 10,* 349–360.

Cole, C. L. (1994). Resisting the canon: Feminist cultural studies, sport, and technologies of the body. In S. Birrell & C. L. Cole (Eds.), *Women, sport, and culture* (pp. 5–29). Champaign, IL: Human Kinetics.

Deegan, M. J. (1985). Multiple minority groups: A case study of physically disabled women. In M. J. Deegan & N. A. Brooks (Eds.), *Women and disability: The double handicap* (pp. 37–55). New Brunswick, NJ: Transaction Books.

Deegan, M. J., & Brooks, N. A. (Eds.). (1985). *Women and disability: The double handicap.* New Brunswick, NJ: Transaction Books.

DePauw, K. P. (1994). A feminist perspective on sports and sports organizations for persons with disabilities. In R. D. Steadward, E. R. Nelson, & G. D. Wheeler (Eds.), *VISTA '93—The Outlook* (pp. 467–477). Edmonton, Alberta: Rick Hansen Centre.

DePauw, K. P. (1997a). Sport and physical activity in the life-cycle of girls and women with disabilities. *Women in Sport and Physical Activity Journal, 6*(2), 225–237.

DePauw, K. P. (1997b). The (in)visibility of disability: Cultural contexts and sporting bodies. *Quest, 49,* 416–430.

Fine, M., & Asch, A. (1985). Disabled women: Sexism without the pedestal. In M. J. Deegan & N. A. Brooks (Eds.), *Women and disability: The double handicap* (pp. 6–22). New Brunswick, NJ: Transaction Books.

Fine, M., & Asch, A. (Eds.). (1988). *Women with disabilities: Essays in psychology, culture, and politics.* Philadelphia: Temple University Press.

Guthrie, S. R. (1997). Defending the self: Martial arts and women's self-esteem. *Women in Sport and Physical Activity Journal, 6*(1), 1–28.

Hall, M. A. (1996). *Feminism and sporting bodies: Essays on theory and practice.* Champaign, IL: Human Kinetics.

Hanna, W. J., & Rogovsky, B. (1993). Women with disabilities: Two handicaps plus. In M. Nagler (Ed.), *Perspectives on disability* (2nd ed., pp. 109–120). Palo Alto, CA: Health Markets Research.

Henderson, K. A., & Bedini, L. A. (1995). "I have a soul that dances like Tina Turner, but my body can't": Physical activity and women with mobility impairments. *Research Quarterly for Exercise and Sport, 66,* 151–161.

Henderson, K. A., Bedini, L. A., & Hecht, L. (1994). "Not just a wheelchair, not just a woman:" Self-identity and leisure. *Therapeutic Recreation Journal, 28*(2), 73–86.

Highlen, P. S., & Finley, H. C. (1996). Doing qualitative analysis. In F. T. L. Leong & J. T. Austin (Eds.), *The psychology research handbook* (pp. 177–192). Thousand Oaks, CA: Sage.

Hopper, C. A. (1986). Socialization of wheelchair athletes. In C. Sherrill (Ed.), *Sport and disabled athletes* (pp. 197–202). Champaign, IL: Human Kinetics.

Horvat, M., French, R., & Henschen, K. (1986). A comparison of the psychological characteristics of male and female able-bodied and wheelchair athletes. *Paraplegia, 24,* 115–122.

Locke, L. F. (1989). Qualitative research as a form of scientific inquiry in sport and physical education. *Research Quarterly for Exercise and Sport, 60,* 1–20.

Lonsdale, S. (1990). *Women and disability: The experience of physical disability among women.* New York: St. Martin's Press.

Markula, P. (1993). Looking good, feeling good: Strengthening mind and body in aerobics. In L. Laine (Ed.), *On the fringes of sport* (pp. 93–99). St. Augustin, Germany: Academia.

Messner, M. A. (1988). Sports and male domination: The female athlete as contested ideological terrain. *Sociology of Sport Journal, 5,* 197–211.

Morris, J. (1993). Gender and disability. In J. Swain, v. Finkelstein, S. French, & M. Oliver (Eds.), *Disabling barriers—Enabling environments* (pp. 85–92). London: Sage.

Nilges, L. M. (1997). Five years of Women in Sport and Physical Activity Journal: A content review. *Women in Sport and Physical Activity Journal, 6*(1), 109–129.

Patrick, D. R., & Bignall, J. E. (1984). Creating the competent self: The case of the wheelchair runner. In J. A. Kotarba & A. Fontana (Eds.), *The existential self in society* (pp. 207–221). Chicago: University of Chicago Press.

Sherrill, C. (1993a). Women with disabilities. In G. L. Cohen (Ed.), *Women in sport: Issues and controversies* (pp. 238–248). Newbury Park, CA: Sage.

Sherrill, C. (1993b). Women with disability, paralympics, and reasoned action contact theory. *Women in Sport and Physical Activity Journal, 2*(2), 51–60.

Strauss, A., & Corbin, J. (1990). *Basics of qualitative research: Grounded theory procedures and techniques.* Newbury Park, CA: Sage.

Taub, D. E., Blinde, E. M., & Greer, K. R. (in press). Stigma management through participation in sport and physical activity: Experiences of male college students with physical disabilities. *Human Relations.*

Theberge, N. (1985). Toward a feminist alternative to sport as a male preserve. *Quest, 37,* 193–202.

Wendell, S. (1989). Toward a feminist theory of disability. *Hypathia, 4,* 104–124.

Whitson, D. (1990). Sport in the social construction of masculinity. In M. A. Messner & D. F. Sabo (Eds.), *Sport, men, and the gender order: Critical feminist perspectives* (pp. 19–29). Champaign, IL: Human Kinetics.

Willis, P. (1994). Women in sport in ideology. In S. Birrell & C. L. Cole (Eds.), *Women, sport, and culture* (pp. 31–45). Champaign, IL: Human Kinetics.

12

Aging and Inequality Based on Age

Older People and the Enterprise Society: Age and Self-Employment Propensities, Work, Employment and Society

Fiona Wilson, James Curran, and Robert A. Blackburn

This article makes a valuable and novel contribution to knowledge. It does this by highlighting the topic of older people, which is a neglected topic. A brief search of a social science database produces, I discovered, just 18 articles on the ageing population published in journals in the period 1995–2002. Almost all of the studies have been based abroad, for example in Japan and Singapore, and are concerned with issues like health, housing and social security. Concerns have been expressed in Britain, particularly in the news media, about mounting demands on welfare, health provision and state pension schemes that growing numbers of older people could cause. My research has shown me that the working age population in the UK is expected to become much older. As the baby boom generations of the mid 1960s age,

---

International Small Business Journal, Feb 2003, v21, i1, p117(3).
© 2003 Woodcock Publications Ltd.

the size of the older age groups will change markedly. By 2026 the number of persons aged 60 and over and living in Britain will reach 17.1 million. This is an increase of nearly 50% on the 11.7 million persons in the same age category in 1996. By 2008 the population of pensioners will exceed the number of children (ONS, 2000). Will this growing group be making a significant contribution towards a growth in enterprise?

Older people are, the article tells us, making a valuable contribution to society in terms of being employed; just under 70% of those aged between 50 and pension age are 'economically active' or 'claimant unemployed'. (This compares with almost 80% among the population as a whole.) As people live longer and medical advances help older people lead active lives, they may welcome early retirement as a time to establish their own business and a means of gaining control over their lives. Middle age can be seen as the beginning of a 30-year period of personal employment and self-indulgence (Scase, 1999), rather than as a time to be a drain on national resources. As the article argues, there is then a positive picture to paint.

Curran and Blackburn also acknowledge that there is a less optimistic picture. Older workers are disproportionately represented among the poorest in society. They are more likely to suffer ill health. They experience discrimination at work (Dibden and Hibbert, 1993). They are discriminated against through voluntary or compulsory redundancy on grounds of age or costs (Parsons and Mayle, 1996). One way of reducing welfare dependency and discrimination against older people in the workplace would then be for more of them to enter self-employment or small business ownership.

Recent UK government initiatives have encouraged older people to remain economically active. Initiatives aimed at getting people back into employment, promoting self-employment and business ownership amongst older people include PRIME, New Deal 50+ scheme and the Employment Zones initiative (DfEE, 2001). Of those aged 65 or over, who are economically active, a quarter are self-employed (Tilsey, 1995). Older people seem to run successful businesses. Cressy and Storey (1995) found that only 19% of start-ups survived after six years, but 70% of businesses, with owner managers over 55, were still in business. Older people are more likely to have the experience and assets for business ownership than younger age groups (Fry, 1997). One would expect older people's motivation for setting up a business to be a strong desire for independence and control. Maybe it is because they face discrimination in employment, but there is a dearth of research.

There is little research on attitudes to or experiences of small business ownership, particularly older people's views. This research by Curran and Blackburn is, as far as I can see, the only recent British research on this topic. They report the findings of a postal survey of 1000 individuals aged 50–75. They found that of the 463 people who responded, 7% were self-employed. Amongst those already retired and not in paid employment, only 7 of the 182 in this category said they would like to be self-employed. Amongst those still in paid employment, just under 15% were attracted to self-employment. There is a wealth of information in the article about the reasons given for and against self-employment from this age group. The article helps challenge stereotypes and look at facts and figures. For example, it was found that a large proportion of respondents were reluctant to opt for self-employment because of reasons related to age, perhaps arising from worries about health and energy levels. Despite policy-makers' promotion of self-employment amongst the older age group, the findings presented here are not encouraging. Entrepreneurship does not appear to be the solution to the 'crisis' of the ageing population. If there is a trend towards older people opting to work for themselves, it appears to be starting slowly from a low base. More needs to be learnt about older age groups and the ways in which they relate to economic activity.

This work by Curran and Blackburn is well informed. It helps us work logically through arguments for and against the likelihood of an increase in enterprise amongst the older age group in society and presents valuable evidence. As a researcher in gender, I would have liked to see an analysis by gender, particularly following on from Goffee and Scase's (1985) intriguing finding that widowed women have been found to exhibit higher entrepreneurial rates than any other category. This is an area of research that needs to be followed up as there has been a substantial increase in the number of mid-life small business start-ups (Fuller, 1994). Many individuals invest redundancy payments or occupational pensions, providing themselves with a business that they envisage will provide them with stable employment until the end of their working life (Fuller, 1994). We need, in my view, to know more.

REFERENCES

Cressy, R. and Storey, D. (1995) *New Firms and Their Banks*. Warwick and London: Centre of Small and Medium Sized Enterprise, Warwick University Business School and NatWest Bank.

DfEE (2001) Action on Age: Report on the Consultation on Age Discrimination in Employment (March). London: DfEE.

Dibden, J. and Hibbett, A. (1993) "Older Workers—an Overview of Recent Research", *Employment Gazette* June: 237–50.

Fry, A. (1997) "Shades of Grey", *Marketing 24* April.

Fuller, T. (1994) (ed.) *Small Business Trends 1994–1998*. Durham: Durham University Business School.

Goffee, R. and Scase, R. (1985) *Women in Charge: the Experience of Female Entrepreneurs*. London: Allen and Unwin.

ONS (1999) *Social Focus on Older People*. London: HMSO for the Office of National Statistics.

ONS (2000) *Population Trends (Spring)*. London: The Stationery Office.

Parsons, D. and Mayle, L. (1996) "Ageism and Work in the EU: A Comparative Review of Corporate Innovation and Practice" (working paper). Cranfield: The Host Consultancy and Cranfield University.

Scase, R. (1999) *Britain Towards 2010: The Changing Business Environment*. London: Department of Trade and Industry.

Tilsey, C. (1995) "Older Workers: Findings from the 1994 Labour Force Survey", *Employment Gazette* April: 133–40. Professor Fiona Wilson University of Glasgow, UK.

13

The Economy and Work in Global Perspective

Long-Term Unemployment Among Young People: The Risk of Social Exclusion

Thomas Kieselbach

The European Union considers long-term unemployment among youth an impediment to their full integration into society. The Commission of the European Union has, therefore, supported research into the mechanisms by which youth unemployment leads to "social exclusion." This paper provides a brief summary of the project's findings. These imply that theories and previous research concerning social exclusion and social support can help explain the effects of youth unemployment although the phenomena vary from country to country. Results also suggest that the construct of social exclusion leads to interventions that reduce the personal and societal costs of youth unemployment.

Unemployment has increased over the last 25 years in all countries of the European Union (EU). The worldwide economic crisis of the mid-70s and early 80s induced the increase since sustained by labour market reforms intended to increase competitiveness in the emerging "global" economy.

As shown in Table I, youth unemployment exceeds total unemployment in all EU countries except Germany where the dual education system and

TABLE I. Overall and Youth (i.e., Under 25 Years of Age) Unemployment Rates for 6 European Countries in 1997

	Germany	Sweden	Belgium	Greece	Italy	Spain
Overall unemployment rate	9.7	10.2	9.5	9.6	12.1	20.9
Youth unemployment rate	10.3	20.9	23	31	33.1	38.8

associated training "spreads" full entry in the labor market over a longer period than elsewhere (Kieselbach, 1998). Researchers in Europe, and elsewhere, report that extended unemployment is a risk factor for behavioral and health problems (Kieselbach, 1988; Olaffson & Svensson, 1986; Spruit & Svensson, 1987; Winefield, Tiggemann, Winefield, & Goldney, 1993).

As part of its response to youth unemployment, the EU commissioned the research project "Youth Unemployment and Social Exclusion: Objective Dimensions, Subjective Experiences, and Innovative Institutional Responses in Six European Countries" (YUSEDER). The principal research objective of the project is to use the theoretical framework of social exclusion to identify mechanisms linking the experience of long-term youth unemployment to various dimensions of social disintegration. The project studies the mechanisms that exacerbate the stress of unemployment (i.e., vulnerability factors) as well as protective mechanisms that prevent or reduce the risk of social exclusion. The project focuses on three Northern European countries (Sweden, Belgium, Germany) and three Southern European countries (Spain, Italy, Greece) (Kieselbach, 2000a, 2000b; Kieselbach, van Heeringen, La Rosa, Lemkow, Sokou, & Starrin, 2001).

THE CONCEPT OF SOCIAL EXCLUSION

Kronauer (1998) described a comprehensive theoretical definition of social exclusion based upon the use of the term in France (Castel, 1994, Paugam, 1996, Silver, 1998) and the concept of underclass in the USA. Kronauer asserts that the ever-increasing unemployment rates have become a permanent social reality in the EU with the consequence that more and more persons with lesser skills cannot lead a life that meets societal standards for material and social well-being (Starrin, Rantakeisu, Forsberg, & Kalander-Blomqvist, 2000).

Social exclusion can be understood not only by focussing on what it means to be excluded versus included, but also on how each of these circumstances either enlarge or diminish the vulnerability of the individual. According to Kronauer, social exclusion arises from the sum and interaction of six types of exclusion.

Labor market exclusion—The economic forces alluded to above create barriers to employment for those with relatively few skills. Failure to enter, or re-entering, the workforce induces feelings of marginality and of being of little value to society.

Economic exclusion—Poverty induced or sustained by labor market exclusion leads to financial dependency on the welfare state and the loss of ability to financially support oneself or one's family at the norm for society.

Institutional exclusion—The poor and unemployed do not have access to private institutions (e.g., banks, insurance companies) that others can to turn for help in reducing the uncertainties of life. Instead, the unemployed must turn to state institutions that serve marginalized persons. This can induce feelings of dependency that lead to shame and passivity.

Social isolation—The circumstances described above lead to loss of, or retreat from, one's social network and the reduction of social contacts.

Cultural exclusion—The inability to live according to the socially accepted norms and values leads to stigma and sanctions from the social surroundings.

Spatial exclusion—All the above circumstances lead to geographic concentration and segregation of persons with limited financial possibilities. They often live in areas with missing infrastructure (e.g., lack of transportation, shops, cultural events).

The YUSEDER project did not intend to contribute to the ongoing debate over the validity of the social exclusion construct, or to the literature concerned with the effect of national welfare policy on exclusion (e.g., Gallie & Paugam, 2000). We, instead, used the construct to organize and convey our record of the experiences of unemployed youth across Europe. The project, moreover, recorded the effect of welfare policy on these youths by gathering their perceptions of the phenomena that such policy supposedly targets. Chief among these are the social and institutional support available to them.

LONG-TERM YOUTH UNEMPLOYMENT AND SOCIAL EXCLUSION

The literature includes arguments that Kronauer's concept of social exclusion describes youth who have suffered long-term unemployment in the EU (Kieselbach, 1997). The YUSEDER Project has attempted to determine if this argument has empirical support. The empirical element of the project is based on 300 interviews (50 interviews in each country). Respondents were between 20–24 years of age and officially registered as unemployed for at least 12 months. The gender distribution of the national samples, which ranged from 54% female in Germany to 66% in Greece, was the same as the distribution among long-term unemployed persons in the country.

Each national sample also had the same proportion of lower versus higher qualified persons as among the long-term unemployed persons within the country. These ranged from 14% with lower qualifications in Greece to 58% in Belgium. "Lower qualifications" refers to dropouts, early school-leavers and persons without further (vocational) training, whereas "higher qualifications" describes persons with high school education and/or a full vocational training. Immigrants were excluded from the study because language barriers made it too difficult to conduct the interviews.

With the exception of Sweden, there were wide disparities in unemployment rates across regions. The national samples were, therefore, drawn from high and low unemployment areas.

All respondents were given the Problem-Focused Interview (PFI) developed at the University of Bremen (Witzel, 1987, 1995). Respondents also completed an additional interview based upon Kronauer's six dimensions of social exclusion (Kronauer, 1998). Items measured several thematic constructs. The first concerned the labor market experiences of the young person. Items dealt with the concrete experience of long-term unemployment; specifically the inability to enter the labor market after school (e.g., structural barriers to getting a job; forms of self-exclusion/resignation).

The second measured construct dealt with the economic situation of the interviewee including constraints resulting from poverty, as well as strategies (e.g., part-time or "irregular" work, working in the unofficial or "submerged" economy) the respondent had developed to deal with them. For the southern European countries additional questions addressed participation in the "submerged" or untaxed and unmeasured economy because there is widespread belief that such economies supplement the opportunities in the "official" economy.

The third construct concerns experiences with institutions such as schools, training institutions, unemployment and social security offices, as well as public and private service institutions. Items also assessed whether these experiences induced dependency, shame, or passivity.

The fourth group of items assessed both the scope and quality of social relationships (i.e., family, partner, friends). The fifth group focused on the cultural norms and sociopolitical experiences of the interviewee.

The sixth set of items concerned the spatial environment both on a structural level (e.g., housing situation, residential area), and the personal experiences of being at home or feeling secure.

The individual cases were assigned to three risk groups of social exclusion: long-term unemployed young people at high risk of social exclusion, increased risk of social exclusion, or who show only a low risk of social exclusion. We use "risk of social exclusion" (indicating a process) rather than "social exclusion" (as a final state).

The group of long-term unemployed young people at high risk of social exclusion includes cases that scored high on at least three aspects of social exclusion. At least two of the three aspects had to be labor market exclusion, economic exclusion, or social isolation (i.e., essential criteria) because these three were the most common experiences across the national samples.

The increased-risk group included two types of respondents. The first scored high on two of the essential but no other criteria. The second included

TABLE II. Percentage Distribution of Respondents by Type of Social Exclusion

	Total of Respondents	High Risk of Social Exclusion	Increased-Risk of Social Exclusion	Low Risk of Social Exclusion
Sweden	49	16%	55%	29%
Belgium	50	46%	26%	28%
Germany	50	48%	32%	20%
Greece	50	12%	40%	48%
Italy	50	16%	18%	66%
Spain	50	26%	50%	24%

those who scored high on one of the central and any number of other aspects. The third group of the sample that met at maximum one nonessential criterion for social exclusion was categorised as at low risk of social exclusion. As shown in Table II, cases at high risk of social exclusion prevail in Belgium and Germany. Most of the young people in the Swedish and Spanish samples fall in the increased-risk category. Youth in Greece and Italy fell mostly in the low-risk category.

As implied by the criteria, a high risk of social exclusion arises from the combination of labor market exclusion, economic exclusion, and social isolation. In the Belgian, German, and Swedish studies, persons in the high-risk group also experienced cultural exclusion. Persons at high-risk in Greece and Italy were also likely to be at risk of spatial exclusion. Due to financial and other problems in their families, high-risk youth tended to be disadvantaged from childhood. They had low qualification level and self-esteem. They tended to be passive and had relatively little support from the social environment and governmental institutions. In Germany, Spain, and Sweden, high-risk youth engaged in deviant behaviour including drug use.

The principal cause of high risk in all countries appeared to be a lack of qualifications needed to escape unemployment. Persons with low qualifications had access only to poor and precarious jobs.

Young unemployed people at high risk of social exclusion also exhibited passivity toward the labor market. Passivity toward the labor market means that high-risk respondents saw few or no chances of finding regular work. They often stopped seeking work, or sought with little effort. Efforts to enhance qualification also waned.

Low social support among high-risk respondents appeared to increase the risk of social exclusion. In Italy, Greece, and Spain, the lack of support by the family appeared most important. In Belgium, Germany, and Sweden a lack of social contacts seemed most important.

Low institutional support characterized all respondents except those in Sweden. Even high-risk unemployed youths in Sweden reported high institutional support.

Finally, high-risk respondents in all countries reported personality traits, such as low self-esteem and poor mental health, that increase the risk of social exclusion. German, Greek, and Spanish respondents in this group also reported the most deviant behaviour including drug dependency.

Andi typifies the high-risk group. Twenty years old, Andi comes from East Germany and has been without work since he left school at the age of 14. He lives alone in a one-room flat and has a girlfriend who is expecting his child. Problems with his family and socialization deficits led to failure in school and, in turn, impeded his integration into the labor market. Other circumstances make him vulnerable to the various dimensions of functional exclusion that can lead to social exclusion. Alcohol and drug consumption and the increased propensity for violence, for example, make it all the more likely he will be socially excluded. The only countervailing forces are a circle of friends and the emotional support of his girlfriend.

Andi exhibits illness that is both a cause and effect of his exclusion. He has asthma and hallucinations, which mainly result from the drug consumption. He also exhibits nervousness, restlessness, and aggressiveness. Furthermore, he shows low self-esteem and expresses high insecurity towards his personal future.

The group of young people at increased-risk of social exclusion seems to be relatively heterogeneous. In all national samples, labor market exclusion was a constant among those at increased-risk. Greeks were most likely to suffer institutional and spatial exclusion as secondary forms of exclusion while Swedes and Spaniards were most likely to suffer economic exclusion.

Persons in the increased-risk group were likely to lack access to educational resources that could improve their qualifications. A high degree of family support (Greece, Italy, Sweden, Spain) or support from the social environment (Belgium, Germany) buffered this group from social exclusion. Furthermore, general activity, which must also be regarded as a protective factor, was at a high level.

The dimension that most frequently made the difference between high and increased-risk of social exclusion was social isolation. The strong link to the family described in the Italian, Greek, and Spanish studies, however, often has a negative aspect as well. Good social networks (especially family, but also friends) may reduce exclusion, but they also induce a feeling of economic dependence.

Rocio belongs to the intermediate group of increased-risk. She is 21 years old and lives with her parents in the south of Spain (Andalucia). Rocio has been unemployed for 2 years. After finishing obligatory school at the age of 15, she looked for work and, after a few months, got a job as a waitress in a tourist hotel. In the following years she has worked intermittently in several jobs, but always without a formal contract. At 19 she participated in a 6-months-course for carpentry from the employment office.

The central vulnerable factors for Rocio are the tense relation with her father and the poor economic situation of her family with her mother working in precarious jobs as well. Her poor educational background and the lack of professional career can also be considered as vulnerability factors. Finally, Rocio lives in a social environment in which many people depend on irregular work and in which petty crime is common. She has psychosocial problems including a strong feeling of guilt due to economic dependence on the family. After a time without work in irregular jobs, she became quite depressed and thought of leaving her parent's home.

A protective factor for Rocio is emotional and economic support, especially from her mother. Rocio assumes that she will always be able to find submerged jobs and will be able to survive in her environment. Another important protective factor is the support from her boyfriend and from friends in a similar social situation.

Long-term unemployed youth at low risk of social exclusion meet at maximum only one nonessential criterion of social exclusion. Respondents in this group regard unemployment as temporary, and a time for personal development and planning. In the Belgian and Italian samples, this group also included young persons not primarily aiming for integration into the working world.

People at low risk of social exclusion feel neither socially isolated nor economically excluded. In southern Europe they receive sufficient financial support from their families and in northern Europe from state institutions. Compared to the groups at higher risk, youth at low risk of social exclusion have higher qualifications, are more actively seeking for a job, are in a financially relatively secure situation, and are supported by their social environment. They are, moreover, satisfied with their benefits from governmental institutions. Many persons of this group are socioculturally active and have high self-esteem.

Support (financial, social, and institutional) appeared to protect the low-risk group from social exclusion. Social support enabled some interviewees to subordinate the search for a job to the maintenance of their current lifestyle, which is characterised by temporary jobs and personal interests like music (Italy). Some respondents in Belgium actually choose to remain unemployed.

Youth at low risk of exclusion usually show protective personality features. All have a high level of self-esteem and good communication skills. They are able to make decisions, to plan out and implement positive changes in their lives, and to cope with new requirements. Youth at low-risk of social exclusion in Sweden were less likely to depend to a lower extent on institutional support due to their willingness and ability to help themselves.

Francesco typifies the low-risk group. He is 24 years old and lives with his parents in the south of Italy (Naples). After school, he started a training course at the university but, after a short time, dropped out, did a course as video-operator and his military service. He has been unemployed for 3 years.

Francesco's main weakness in the labor market is his poor training. He has done several irregular jobs in the past and intends to work in future in the submerged economy. Francesco is protected from the risk of social exclusion by a number of factors including strong support from friends and family in an integrated neighbourhood, his resourcefulness and flexibility, and his strong desire to receive more professional training. In regard to his unemployment situation, he can be described as a proactive person. From the psychosocial point of view, Francesco is neither ashamed of his situation nor does he appear to be discouraged about his personal future. He does not rely on the Employment Office to find odd jobs. He is well integrated into the neighbourhood. Francesco does not appear dissatisfied with being forced to prolong his adolescence by living with, and by being economically dependent on, his parents.

High-risk respondents from all participating countries except Greece reported psychosocial strains resulting from unemployment. The level of strain and its effects, however, appear to differ across the participating countries. The level of psychosocial strain is greater in the northern European countries than in Greece, Italy, and Spain. In Greece, health problems among high-risk unemployed young people were more likely to preexist than result from unemployment. In Spain, social support seemed the best predictor of the effects of strain. High-risk respondents with little social support were most likely to feel lonely, bored, afraid of the future, and to suffer from feelings of depression.

High-risk Northern European respondents reported a multitude of psychosocial stressors and effects. Belgians, Germans, and Swedes were most likely to report financial stress, fear of the future, missing a perspective for life, and feelings of dependency.

We found that long-term unemployed youth at high risk of social exclusion face multiple health problems. Selection of ill persons into the high-risk group seems to vary by country. The evidence for a causal effect of unemployment on self-esteem, which was characteristically low among high-risk youth in all countries, seems strongest, for example, in Belgium, Germany, and Spain. Suicidal ideation, on the other hand, seems to increase as an effect of unemployment only in Germany. In Greece and Italy no negative effects of unemployment on health were observed.

In all six studies the health of those at increased-risk of exclusion is better than of youth at high risk of exclusion. Possible effects of exclusion were, however, also detected in this intermediate group. Psychosocial strain was commonly reported in this group, although a lower level than reported among youth at high risk of exclusion. Youth in the intermediate group seemed to be protected from health effects of strain by support from their parents. Young Italians in the increased-risk group, who also had parental support, apparently had very few strains and fewer health effects. The only strains they reported were a lack of control of their situation, and a lack of structure of their everyday life.

The health of intermediate-risk Greek youth appeared to be largely independent of their vocational status. In Northern Europe, youth at increased-risk of exclusion were in a slightly better health than youth at high risk of exclusion. Compared to Southern European studies, important psychosocial stressors included financial burden (Germany, Sweden) and financial insecurity (Belgium).

Respondents at low risk of social exclusion were in the best health in all countries. Low-risk youth report being relatively free of strain. They describe themselves as active and having high self-esteem and good communication skills. No case of risky health-related behaviour is reported. In all three Northern European countries, social support and financial protection appear equally important protective factors for this group. High qualification clearly protected those in Belgium and Germany. Active behaviour can be observed in the reports from Sweden and Germany. For some young people with low risk of social exclusion unemployment was a deliberate choice (Belgium, Italy), some use their unemployment as a time for orientation (Sweden, Germany). Thus, they assess the strain resulting from unemployment as being rather low.

INNOVATIVE INSTITUTIONAL RESPONSES

The third phase of the YUSEDER research project focused on the institutional responses to youth unemployment and social exclusion in the participating countries. Every young person who has been out of work for at least 100 days in Sweden, for example, receives, by law, an offer of employment or training. The Swedes view training of unemployed young people as the key to reducing youth unemployment. Projects and initiatives that focus on individual job counselling and supervision for unemployed young people are emphasized. The object is to introduce young people to work through practical experience.

The German government provides 1 billion Euros (in addition to the funds for regular labor market programmes), for projects to promote young people's integration into the labor market and to minimize the risk of social exclusion. These efforts have three characteristics: 1) training and qualification, 2) cooperation and networking of the various mediators and organisations, and 3) individual psychosocial stabilization and personality development of those affected. Here, we must especially point out projects and programmes that attempt to set up training and work positions with the help of placement agencies mediating between businesses and the unemployed young people. In such cases, it is of critical importance that an individual developmental plan be designed in conjunction with the young people themselves and that supportive supervisors monitor progress on the plan.

In Belgium, concrete actions against youth unemployment and social exclusion have been undertaken primarily on the regional level. The regional efforts have in common that personal support is viewed as the most important component for success.

Reports from Spain emphasize that projects concentrating merely on training unemployed young people fall short. Successful programs are those in which elemental capabilities and qualifications are taught and young people are encouraged to achieve more independence and self-assurance in structuring their vocational future.

In contrast, less national government influence could be found in Italy and Greece. The European Social Fund, however, supports many projects and initiatives in these countries.

In general, innovative approaches toward avoiding youth unemployment and the risks of social exclusion take into account that training and qualification alone are not enough to guarantee a longer-term integration of unemployed young people. Measures are required in which the young person first receives help with their personal life situation and emotional development. Only after a phase of personality stabilisation and improvement in their social situation do efforts toward integration into the labor market seem meaningful.

EU Employment Guidelines provide the framework and concrete directives for the annual National Action Plans for Employment in each of the 15-member countries. Since 2001, the governments of the member countries are also obliged to deliver an Annual National Action Plan for Social Inclusion in which they have to report on concrete measures. While the EU cannot dictate the substance of these plans, it intends to monitor social inclusion in the member states through its "Targeted Socio-Economic Research" (TSER) program. The YUSEDER project is part of this program.

CONCLUSIONS

Unemployment threatens the overall integration of young people into society. The most important vulnerability factors that contribute to an increase of the risk of social exclusion for young unemployed people in the long-term are in low qualification, passivity in the labor market, a precarious financial situation, low or missing social support, and insufficient or nonexistent institutional support. The most important protective factor for unemployed youth is social support. While integration into social networks is of great importance for youth from Northern Europe, in Southern Europe the family is more important. Especially due to the high level of family support, the number of youth at high risk of social exclusion appears lower in Southern compared to Northern Europe.

Social origin can be a protective factor for the youth as well as a decisive vulnerability factor: Poverty and other social problems in the family can increase the risk of social exclusion for the youth. This can be interpreted in the sense that the effects of social origin are reinforced by the experience of long-term unemployment of young people. The higher involvement of young people in the Southern European countries in irregular work (81% vs. 24%) acts as a buffer and as a trap (Borghi & Kieselbach, 2000, 2001). Bridges should be found that help youth out of irregular work and into the labor market without stigmatising participation in the submerged economy. To reduce the risk of social exclusion and to enhance controllability of the future for young unemployed people a concept like the Swedish "100 days guarantee" seems to be innovative and exemplary. This includes the offer of employment or training from the side of institutions at the latest after 100 days of being out of work.

The long running debate in the EU over whether and how labor market policies can reduce the psychosocial strain associated with unemployment has converged on the concept of sustainable employability as one pillar of the EU Employment Guidelines. To avoid a reductionist, individualizing approach to implementing this policy, action has been conceived of as a balance between enhancing personal resources (e.g., competencies and qualifications, personal initiative, and entrepreneurship) and providing more social resources (e.g., social support during occupational transitions). Thus implementation strategies include an individual dimension as well as a social dimension requiring changes in the work organization, the school system, training system, and retraining system.

In the case of unemployment, early professional counselling may reduce the risk of social exclusion by preventing the psychosocial reactions that lessen success in the labor market.

Another intervention suggested by our findings is the creation of programs that keep first-time job loss from turning into long-term unemployment. Here the concept of a "social guarantee" comparable to the Swedish example mentioned above may give the unemployed person fixed-term employment, or at least the offer to participate in a comprehensive training scheme. This guarantee can be timed such that young workers are offered the employment or training after relatively short periods of unemployment.

REFERENCES

Beelmann, G., & Kieselbach, T. (2001). Finding good practices for interventions to combat long-term youth unemployment. In S. Mannila, M. Alakauhaluoma, & S. Valjakka (Eds.), *Good practice in finding good practice: International Workshop on Evaluation* (pp. 85–96). Helsinki: Rehabilitation Foundation.

Borghi, V., & Kieselbach, T. (2000). The submerged economy as a trap and a buffer. Comparative evidence on long-term youth unemployment and the risk of social exclusion in Southern and Northern Europe. EU-Workshop on Unemployment, Work and Welfare. European Commission (DG Research), Brussels, 9–11 November, 2000.

Borghi, V., & Kieselbach, T. (2001, September). L'economia sommersa, trappola o risorsa?. [Submerged economy, a trap or a resource?] *Impresa Sociale, 59,* 21–31.

Castel, R. (1994). De l'indigence a l'exclusion: La desaffiliation [From poverty to exclusion: Disaffiliation]. In J. Donzelot (Ed.), *Face a l'exclusion: Le modele francais* [In the face of exclusion: The French Model]. Paris: Edition Esprit.

Eurostat. (1998). *Basic statistics of the community.* Bruxelles: Author.

Gallie, D., & Paugam, S. (2000). *Welfare regimes and the experience of unemployment in Europe.* Oxford: Oxford University Press.

Kieselbach, T. (1988). Youth unemployment and health effects. *The International Journal of Social Psychiatry, 34*(2), 83–96.

Kieselbach, T. (1998). Arbeitslosigkeit [Unemployment]. In Statistisches Bundesamt [National Institute for Statistics in Germany] (Ed.), *Gesundheitsbericht fur Deutschland* [Health Report Germany] (pp. 116–121). Stuttgart, Germany: Metzler-Poeschl.

Kieselbach, T. (Ed.). (2000a). Psychology of social inequality: Vol. 9. Youth unemployment and health: A comparison of six European countries (YUSEDER Publication No. 1). Opladen: Leske & Budrich.

Kieselbach, T. (Ed.). (2000b). Psychology of social inequality: Vol. 10. Youth unemployment and social exclusion: A comparison of six European countries (YUSEDER Publication No. 2). Opladen: Leske & Budrich.

Kieselbach, T., van Heeringen, K., La Rosa, M., Lemkow, L., Sokou, K., & Starrin, B. (Eds.). (2001). Psychology of social inequality: Vol. 11. Living on the edge: A comparative study on long-term youth unemployment and social exclusion in Europe (YUSEDER Publication No. 3). Opladen: Leske & Budrich.

Kronauer, M. (1998). "Social exclusion" and "underclass"—New concepts for the analysis of poverty. In H.-J. Andress (Ed.), *Empirical poverty research in a comparative perspective* (pp. 51–75). Aldershot, England: Ashgate.

Olaffson, O., & Svensson, P.-G. (1986). Unemployment-related lifestyle changes and health disturbances in adolescents and children in the Western countries. *Social Science and Medicine. 22*(11), 1105–1113.

Paugam, S. (1996). *A new social contract? Poverty and social exclusion: A sociological view.* EUI Working Papers, European University Institute, No. 96/37.

Silver, H. (1998). Policies to reinforce social cohesion in Europe. In A. De Haan & J. Burle (Eds.), *Social exclusion: An ILO perspective* (pp. 2–41). Geneva: International Institute.

Spruit, I. P., & Svensson, P.-G. (1987). Young and unemployed: Special problems. In D. Schwefel & P.-G. Svensson (Eds.), *Unemployment, social vulnerability, and health in Europe* (pp. 196–210). Berlin: Springer.

Starrin, B., Rantakeisu, U., Forsberg, E., & Kalauder-Blomqvist, M. (2000). International debate on social exclusion. In T. Kieselbach (Ed.), *Youth unemployment and social exclusion. A comparison of six European countries* (pp. 15–25). Opladen: Leske & Budrich.

Winefield, A. H., Tiggemann, M., Winefield, H. R., & Goldney, R. D. (1993). *Growing up with unemployment. A longitudinal study of its psychological impact.* London: Routledge.

Witzel, A. (1987). Das problemzentrierte Interview [The problem-focused interview]. In G. Juttemann (Ed.), *Qualitative Forschung in der Psychologie* [Qualitative research in psychology] (pp. 227–256). Heidelberg, Germany: Asanger.

Witzel, A. (1995). Auswertung problemzentrierter Interviews: Grundlagen und Erfahrungen [Analysis of problem-focused interviews: Fundamental principles and experiences]. In R. Strobl & A. Bottger (Eds.), *Wahre Geschichten? Zu Theorie und Praxis qualitativer Interviews* [True stories? Theory and practice of qualitative interviews] (pp. 49–76). Baden-Baden, Germany: Nomos.

14

Politics and Government in Global Perspective

Mandatory Co-Operation: Because Water Knows No Political Boundaries, Water-Sharing Nations Must Work Together

Jennifer Pedersen and Evan D. G. Fraser

Resource scarcity breeds conflict—or so the news seems to tell us. In October 2002, Israel and Lebanon came close to exchanging gunfire over a Lebanese plan to divert water from the Wazzani River, one of the head-waters of the Jordan River and an important source of fresh water for Israel.

In 2000–2001, the United States and Mexico exchanged heated words over water in the Rio Grande/Rio Bravo Basin, which straddles Texas and Northern Mexico. The basin supports a lucrative export market in alfalfa, melons and chile peppers on the Mexican side, while providing irrigation for 40 crops, including citrus and sugar cane, on the US side of the border. The Mexicans claimed that drought had reduced aquifers to 20 percent of historic capacity while American farmers countered that Mexican mismanagement exacerbated the problem.

Tension does emerge between countries over shared watersheds, usually as a result of disagreement over who has the right to use the water in question, dam construction, failure to take responsibility for water pollution, or when there is not enough water to meet demand. But there has not been an actual war over water for at least 4500 years. Moreover, a growing body of research shows that under certain circumstances, transborder water management can become a source of co-operation between nations. (All figures referenced can be found within the online version of this article, at http://www.infotrac-college.com.)

The International Joint Commission, established in 1909 to prevent water disputes between Canada and the US, is a North American example of successful international water co-operation. On the other side of the world, the Mekong River Commission works to ensure that member nations co-operate in all fields of sustainable development, including the utilization, management and conservation of water. Although the commission comprises only four of the six countries in the Mekong Basin (Laos, Cambodia, Thailand and Vietnam), it maintains a dialogue with Burma and China, thereby providing an important diplomatic forum in a developing region with sometimes strained relations.

Water scarcity is not necessarily a destabilizing force. But political stability does seem to be a prerequisite to transborder water co-operation. In the late 1990s, finally recovering from decades of civil war, Mozambique experienced severe flooding that was partly caused by extreme weather and partly by upstream mismanagement in neighbouring countries. Due to the war, Mozambique had been unable to enter into water management agreements with neighbouring countries—and the impact of this isolation was devastating. According to the Red Cross, flooding killed thousands and stranded up to 100,000 people, destroyed essential infrastructure and redistributed land mines to already-cleared areas. In 1999, South Africa, Swaziland and Mozambique broke this log-jam by beginning transborder scientific work on the Incomati River as part of a newly created Shared Rivers Initiative. Although these scientific benchmarking exercises did nothing to prevent recent flooding, they are a first step in developing appropriate transborder management.

Another important necessity for co-operation is that successful institutions include formal mechanisms to settle disputes between governments. For example, the International Joint Commission acts as an advisor to both the Canadian and US governments on all issues related to water management in shared basins, and provides a forum where transborder disputes can be reconciled peacefully.

In 1944, the US and Mexico signed a water management agreement that obliges Mexico to release water for American agriculture. This agreement provided a forum that allowed the presidents of the US and Mexico to meet in the spring of 2001 to resolve the water problem in the Rio Grande/Rio

Bravo basin. The Governor of Texas described this as a "first step toward a renewed commitment by both countries to allocate and use the valuable water resources of the Rio Grande in a responsible manner." The key seems to be to establish the institutional framework to act proactively to settle disputes before they erupt.

As water scarcity becomes an even more pressing issue due to growing populations, pollution, climate change and industrial demand, competition for scarce resources and conflicts between countries that share water may increase. Water conflicts have the potential to undermine peace and security in many regions around the world. Political will, creativity and inclusive participation—as demonstrated by organizations like the International Joint Commission and the Mekong River Commission—can lead to solutions that will reduce these tensions. Poorly managed, however, water can become an excuse for a war.

FOLLOW UP

The Global Water Partnership promotes co-operation on water resources: www.gwpforum.org/

For information on conflict, see the global chronology of water conflict at The World's Water: www.worldwater.org/conflict.htm

15
Families and Intimate Relationships

Is Posthumous Semen Retrieval Ethically Permissible?

R. D. Ordd and M. Siegler

It is possible to retrieve viable sperm from a dying man or from a recently dead body. This sperm can be frozen for later use by his wife or partner to produce his genetic offspring. But the technical feasibility alone does not morally justify such an endeavour. Posthumous semen retrieval raises questions about consent, the respectful treatment of the dead body, and the welfare of the child to be.

We present two cases, discuss these three issues, and conclude that such requests should generally not be honoured unless there is convincing evidence that the dead man would want his widow to carry and bear his child. Even with consent, the welfare of the potential child must be considered.

There have been sporadic reports of babies born after post-humous conception since the technology became available 50 years ago. Most commonly, a young man has an illness which threatens his fertility or his life—for example, testicular carcinoma. He has some of his semen frozen in order to impregnate his wife in case he should become sterile or to impregnate his widow if he should not survive. In these relatively uncommon cases of posthumous conception, legal questions have been raised about inheritance and eligibility for survivor benefits. (1) Few questions have been raised, however, about the ethics of the procedure because the semen was donated voluntarily, before death, with the expressed intent of use after death.

Retrieval of viable sperm after death, first described by Rothman in 1980, (2) raises significantly different issues. It has been reported in the popular press that a baby has been born using posthumous sperm collection (3) after a young man died unexpectedly from an allergic reaction. At his wife's request, sperm was collected 30 hours after death. Fifteen months later his sperm were used to impregnate his widow.

Such requests are infrequent; 82 were reported in the US in a 1997 study, of which about one-third were honoured. (4) Reported successes will likely encourage more requests. In addition, the advent of intracytoplasmic sperm injection (ICSI) now makes it possible to fertilise an egg in the laboratory using a single sperm rather than the several cubic centimetres of semen required for artificial insemination. After describing the technical feasibility of sperm retrieval after death, however, a standard textbook of urology concludes: "[t]he ethical appropriateness of such retrieval is the most important issue surrounding its use". (5)

An identical endpoint—the dramatic birth of a dead man's baby—makes voluntary sperm donation before death and involuntary sperm retrieval after death seem only a small step apart. The difference between these two procedures is not, however, a small step.

In Western society, there is no universal prohibition of post-humous gamete retrieval or posthumous in vitro fertilisation. However, recently reported successes have prompted discussion in the popular press. (6) These practices raise at least three significant ethical questions. First, the method of sperm collection raises issues about respectful treatment of a dead body. Second, there is the issue of consent, important in all invasive procedures. Third is the issue of the welfare of the child to be. We will present two cases which highlight these issues.

CASE 1

A 28 year old man had been married for six years and he and his wife were childless. He became depressed after a marital separation three months ago. Two weeks ago he started antidepressant medication, but today he was brought to the hospital by paramedics with a self inflicted gunshot wound to his head. Six hours later he was pronounced dead in the intensive care unit. He is the only child of his parents. Just before he died, they asked the intensive care physician to arrange for sperm retrieval after his death so that they might have a grandchild. They said they are certain he would want his biological line continued, and they thought his estranged wife, who had been contemplating reconciliation, would be willing to have his child.

A clinical ethics consultant was asked to review the situation and make recommendations. After talking with the parents and widow, he was unable to elicit any substantiating evidence that the man would want his widow to bear his child. He recommended against the requested sperm retrieval.

CASE 2

A 36 year old previously healthy man was admitted with pneumonia. He developed adult respiratory distress syndrome requiring assisted ventilation. After 14 days of aggressive treatment, he became obtunded and developed multiorgan system failure, and his wife was informed that he would not survive. She asked if semen could be collected so that she might yet have his child. An ethics consultation was requested.

They had been trying unsuccessfully to have a child for over 10 years. Two months before this illness they saw an infertility specialist and were to begin in vitro fertilisation with her next menstrual cycle. Although this history indicated his desire to become a father, this alone could not be construed as consent for either sperm collection in this circumstance of impending death or for posthumous collection. The uncertainty of whether he would want his wife to be a single mother after his death was troublesome, and his views on the wellbeing of a child raised by a single parent were likewise unknown. The wife believed that he would want this, but they had never discussed the possibility. This presumption was supported by his sister who had talked with him about his intense desire to have children in order to continue his family name. But is a wife's intense desire for her husband's offspring morally relevant, and if it is, is it sufficient to justify the removal of semen without his explicit consent? His physicians, nurses and ethics consultant believed the available information adequately supported his wife's expression of his presumed wishes. Within one hour of his death, his epididymides were removed and frozen.

RESPECT FOR THE DECEASED PERSON

Metaphysically, the person disappears from his or her body at death, but the dead body continues to command respect. (7) This nearly universal respect for the dead body can be observed as the evening news brings images of grieving survivors searching for the bodies of their loved ones who have been lost at the scene of natural disasters around the world. In most cultures, there seems to be an innate drive to recover bodies so they may be given proper burial. Though individuals in some cultures may believe that organs and physical structures of the once living are no longer important, this is distinctly uncommon in Western society. At the same time, this almost sacred respect for the dead body is not held to be absolute. Most people in Western society accept that there are some exceptions when the body may be disturbed before being buried—for example, for postmortem examinations, and for organ or tissue retrieval for transplantation. Other uses of the dead body have led to considerable controversy—for example, the practising of medical procedures by medical trainees.

Postmortem examination has been practised at least since the time of Julius Caesar (8) in order to learn the cause of death, to further understand the pathology or pathophysiology of disease, or for medicolegal reasons. While many individuals still have a natural revulsion to the idea of cutting, opening, and inspecting the dead body, the potential benefits to the medical profession, the family, or to society as a whole have generally overcome this resistance as long as the autopsy procedure is carried out with the maximum possible respect for the departed person.

For over 30 years, after informed consent by all parties, organs and tissue have been retrieved from recently dead bodies and have been used to save thousands of lives. The concept of death using neurologic criteria, developed primarily to allow the timely retrieval of usable organs, has not, however, been universally accepted, (9) and continues to be the subject of controversy. (10) The drive to overcome the current shortage of organs for transplant has led to the development of new techniques for retrieval of organs from "non-heart-beating cadaver donors", (11, 12) with not-unexpected criticism. (13) But overall, a majority of individuals in Western society believe the good achieved by the donation of organs and tissues outweighs initial concerns about the desecration of the dead body. (14) In spite of this consensus, there has been some aesthetic, cultural, and religious resistance to the practice of organ retrieval and transplantation as an enterprise. In addition, some who accept organ transplantation have specific reservations about the disrespectful treatment of dead bodies in some circumstance. For example, Frader has criticised the practice of providing artificial support for a pregnant corpse in order to bring the gestating fetus to viability, maintaining that this represents a profound disrespect for the dead body. (15)

The responsibility for disposition of the dead body has traditionally been given to the family, or when no family is available, to the church or the state. Consent is almost always sought from the family or the state before doing procedures which would otherwise be deemed disrespectful. While occasionally a medicolegal postmortem examination is authorised by the state over the objection of family, most autopsies are preceded by the consent of the family. Likewise, organs are not removed for transplantation without the consent of the family. It is an interesting commentary on contemporary society that even when a person has specifically documented in writing that he or she wants to be an organ donor, transplant teams are unwilling to retrieve organs without explicit agreement from the family. At least in this situation, the wishes of the family are honoured over the explicit wishes of the deceased, perhaps out of concern for liability. But the reverse is not true. If a patient had specifically declined to be an organ donor, transplant teams are unwilling to retrieve organs after his death even with an impassioned plea from his family.

The issues of utility and consent have also dominated discussion of practising medical techniques on newly dead bodies. While a strong case has been made for the utility of such an approach, (16) it has been called "unlawful and unethical" if it is done without family consent. (17) This example of treating dead bodies in less than a respectful way has often been carried out in secret and has clearly not achieved societal acceptance as have autopsy and organ retrieval. (18)

The majority acceptance of some instances of trespassing the integrity of a dead body in order to benefit others indicates that the strong societal mandate to show respect for a dead body is not inviolable. The practice of retrieving sperm from men in coma or recently dead has not, however, been similarly accepted. This practice has been criticised as "perilously close to rape" by law professor Andrews. (19)

CONSENT

The ethical concept of valid consent and the legal doctrine of informed consent have become firmly established as foundational in the practice of modern medicine. (20) Ethically valid consent has three components: (1) the patient must have decision making capacity; (2) he must be given adequate information, and (3) then he must give voluntary consent without coercion.

When a patient does not have decisional capacity, consent may be obtained from a proxy. The proxy's "substituted judgment" ought to reflect the decision that the patient would make if able, based on a written advance directive, the patient's previously expressed wishes, or an understanding of his or her values.

In some situations "implied consent" may substitute for a formal consent discussion. Implied consent may sometimes be inferred from the patient's actions. (21) For example, when a man comes to the emergency room (ER) complaining of chest pain and collapses, it can be assumed he wanted treatment. Different still is "presumed consent" which does not depend on a patient's words or actions, but is based on a theory of human goods. (22) It may be presumed that a person unconscious from injuries sustained in a motor vehicle accident would want to be treated. Thus, when substituted judgment is not possible—for example, in a child who has not developed decision making capacity or in an adult who has not made his wishes known, the proxy is allowed to use the lower and more ill defined standard of "best interests".

When an emotionally involved third party requests sperm retrieval after death, it might seem desirable to seek the same level of certainty we attempt when making other medical decisions, such as limitation of treatment for patients near the end of life. We could use the same hierarchy of (a) patient's current statement; (b) written advance directive; (c) report of previously stated wishes; (d) recognised values, and (e) presumed best interests. When making limitation of treatment decisions, professionals often experience greater discomfort as we move down this scale of increasing uncertainty, but we cannot avoid making the decisions. We must make the best decision possible in the face of limited information and a particular set of clinical circumstances.

This hierarchy, complex as it is to apply in limitation of treatment decisions, may be even less useful in decisions about sperm collection after death. It is rare for a healthy young man to anticipate life-threatening illness, and even more rare for him to contemplate or discuss whether he would want his sperm to be collected after death so that his widow could bear his child. In addition, such a decision, like many end-of-life decisions, is not just about his life. It has major implications for his wife's future and for the future of his potential progeny.

The legal doctrine of informed consent is based on the ethical principle of autonomy. But this right to self determination should not be misinterpreted to mean that whatever the patient wants should be done. Autonomy is a bounded liberty. Though the patient's negative right to be left alone is nearly absolute, the positive right to have what one wants is clearly not absolute. While a patient may request any treatment desired or imagined, the physician, also an autonomous moral agent, is free to decline a treatment he or she believes is not medically indicated, or is felt to be not in the patient's best interests. A patient's request to forgo or stop dialysis when he finds it disproportionately burdensome should almost always be honoured. On the other hand, a request for narcotics to treat chronic tension headaches should not be honoured if the physician believes an alternative treatment is more appropriate.

WELFARE OF THE CHILD-TO-BE

This recognition that the physician has an obligation to do only beneficial procedures and to decline those which are potentially harmful raises the question "who is the patient in posthumous sperm collection"? Does the physician also have a responsibility to decline procedures which may be harmful to a future individual or future generation?

The Human Embryology and Fertilisation Authority of Great Britain requires physicians who provide assisted reproductive technology services to consider the welfare of the potential child before making a decision to proceed. (23) Most physicians would decline to do artificial insemination for a woman who carries a dominant gene for a lethal condition. Some decline to provide services to single women based on studies showing children of single parents do not do as well as children with both parents. (24)

A decision to participate (or not) in helping a woman achieve a pregnancy using the semen of her deceased partner, whether voluntarily frozen for that purpose before death or retrieved posthumously, should consider the welfare of the future child. This calculation is exceedingly difficult, and the conclusion may vary depending on the social circumstances and on personal values. But the issue of the child's welfare cannot be overlooked.

LEGAL ISSUES

The development of new technology often raises ethical questions about it use. Sometimes these "should we . . . ?" questions seem to be settled by statutory or case law, but usually only after an extended time of legal uncertainty. For example, death defined by by neurologic criteria was first proposed in 1968, but settlement of the legal uncertainties did not begin in the US until the proposal for a Uniform Determination of Death Act in 1981. While legislative or judicial determinations often give an imprimatur to a particular action, this does not always fully answer the ethical questions.

There has been some legislative and judicial activity on issues of the status of frozen embryos, parentage after the use of anonymous or designated donated sperm, inheritance after posthumous conception, and other related issues. According to a recent review, however, there have been no laws or cases which give clear guidance about posthumous sperm collection. (25) Based on existing standards of consent, the authors conclude that spousal requests for sperm collection after death should be declined unless there is prior consent or known wishes of the decedent. Their interpretation of the legal climate focuses on the intent of the man, but does not address the issues of treatment of the dead body or the well-being of the potential child.

DISCUSSION

How should we view a request for sperm collection after death? Does it resemble the family's right to give permission for procedures after death such as autopsy, organ donation, and practising medical technology? If so, can we honour family requests for this procedure? Or might the welfare of the potential child be an overriding consideration?

Although the sperm retrieval procedure itself is far less invasive, destructive, or disfiguring than is an autopsy, the invasiveness seems less important than the man's preferences and the long term consequences for the woman and the child. Autopsy and organ retrieval have more immediate consequence to the dead body, but very little ongoing consequence to the deceased or his family. But sperm retrieval has major consequences for his family and also for his own legacy. In our view, there is a difference in kind between autopsy and organ retrieval on the one hand, and sperm retrieval. Giving consent for autopsy or for organ retrieval for transplantation is giving to benefit others. But requesting sperm retrieval after death without the consent of the dead man is not the same; in fact it is not giving at all—it is instead taking, because its aim is to benefit the person making the request. While retrieval of organs after death without the explicit consent of the decedent is likewise taking, it is different in that the family who is giving consent is altruistically giving the organs for someone else's benefit. The parents or woman who request sperm retrieval after death without the explicit consent of the dead man are making a request for their own benefit. Thus, proxy "consent" in this situation is not consent at all.

In our view, if a man had steadfastly refused to have a child while alive, it would be ethically wrong to honour a request to retrieve his sperm for use after his death. At the other extreme, if we had a clear written or verbal statement from him that he would want to father a child after his death, it might be justifiable to assist this endeavour. If, however, as will likely be the situation in most cases, we do not know his wishes, we must rely on the best available information. In our view, it would usually be appropriate to decline such requests. This stance of non-retrieval without the patient's prior consent or known wishes is supported by the American Society of Reproductive Medicine. (26) They go on to say that "such requests pose judgmental questions that should be answered within the context of the individual circumstances and applicable state laws". While this decision might intensify the grief of the widow, and the poignancy of this refusal would seem to heighten the tragedy of his death, it is the ethically most defensible position based on the presumed rights of the dead or dying patient.

Even with consent, how strongly should we consider a man's stated desire to produce offspring or preserve his family name? While the strength of this desire is clearly evident in many discussions of infertility, it is also true that the desires of many infertile couples can be met through adoption. Thus, the use of requested technology is not always needed to satisfy such desires, and some would say the availability of such alternatives make the use of technology unjustified.

In case 1 above, the lack of consent and lack of knowledge of the man's wishes led appropriately to a refusal to comply with the request. In case 2, there was likewise no consent. His willingness to undergo infertility testing and their plan to pursue in vitro fertilisation suggests that this man had a strong desire to have a child. While this evidence gave some guidance to his medical professionals, it provided no indication of his wishes about his wife having his child after his death. Although she was probably in a position to know his wishes better than anyone else, her own self interest could have clouded her understanding of what his wishes would have been in circumstances that he never discussed and probably never contemplated. His sister's statement lent some support to his wife's contention, but this is still not as definitive as if he had made an explicit statement. The decision to honour her request was thus not clear cut, but was a marginal judgment call.

CONCLUSION

A request for sperm retrieval after death should not be honoured unless there is convincing evidence that the dead man would want his widow to carry and bear his posthumously conceived offspring. Even when consent is available, professionals should also consider the welfare of the potential child. The evidentiary standards for such a decision are difficult to define and far from clear.

REFERENCES

1. Banks GJ. Traditional concepts and nontraditional conceptions: Social Security survivor's benefits for posthumously conceived children. *Loyola of Los Angeles Law Review* 1999;32:251–379.
2. Rothman CM. A method for obtaining viable sperm in the postmortem state. Fertility and Sterility 1980;34:512.
3. Baby is born using sperm from baby's dead father. *Los Angeles Times* 1999 Mar 27:4.
4. Kerr SM, Caplan A, Polin G, et al. Postmortem sperm procurement. *Journal of Urology* 1997;167:2154–6.
5. Goldstein M. Surgical management of male infertility and other scrotal disorders. In: Walsh PC, Retnik AB, Vaughan ED, et al, eds. *Campbell's urology* [7th ed]. Philadelphia: WB Saunders Company, 1998:1363.
6. Andrews LB. The sperminator. *The New York Times Magazine* 1999 Mar 28:62–5.
7. May W. Attitudes toward the newly dead. *Hastings Center Report* 1973;1:3–13.
8. Iserson KV. *Death to dust.* Tuscon, AZ: Galen Press, 1994:138.
9. Kimura R. Japan's dilemma with the definition of death. *Kennedy Institute of Ethics Journal* 1991;1:123–31.
10. Shewmon DA. Brainstem death, brain death, and death: a critical re-evaluation of the purported equivalence. *Issues in Law & Medicine* 1998;14: 125–45.
11. Management of terminally ill patients who may became organ donors after death. *Kennedy Institute of Ethics Journal* 1993;3:A1–A15. This paper appeared in The University of Pittsburgh Medical Center policy and procedure manual, which was preprinted as an appendix to this issue of the journal.
12. Orr RD, Gundry SR, Bailey LL. Reanimation: overcoming objections and obstacles to organ retrieval from non-heart-beating cadaver donors. *Journal of Medical Ethics* 1997;23:7–11.

13. Fox R. "An ignoble form of cannibalism": reflections on the Pittsburgh protocol for procuring organs from non-heart-beating cadavers. *Kennedy Institute of Ethics Journal* 1993;3:231–9.
14. Cohen C. The case for presumed consent to transplant human organs after death. *Transplantation Proceedings* 1992;XXIV:2168–72.
15. Frader JE. Have we lost our senses? Problems with maintaining brain-dead bodies carrying fetuses. *Journal of Clinical Ethics* 1993;4:347–8.
16. Iserson KV. Law versus life: the ethical imperative to practise and teach using the newly dead emergency department patient. *Annals of Emergency Medicine* 1995;25:91–4.
17. Goldblatt AD. *Don't ask, don't tell: practising minimally invasive resuscitation techniques on the newly dead.* Annals of Emergency Medicine 1995;25:86–90.
18. Burns JP, Reardon FE, Truog RD. Using newly deceased patients to teach resuscitation procedures. *New England Journal of Medicine* 1994;331:1652–5.
19. See reference 6:65.
20. Applebaum PS, Lidz CW, Meisel A. *Informed consent: legal theory and clinical practice.* New York: Oxford University Press, 1987.
21. Childress JF. Who should decide: paternalism in health care. New York: Oxford University Press, 1982:82–3.
22. See reference 21:83–5.
23. The Human Fertilisation and Embryology Act 1990 (England): ch 37.
24. Gringlas M, Weinraub M. The more things change . . . single parenting revisited. *Journal of Family Issues* 1995;16:29-52.
25. Kahan SE, Seftel AD, Resnick MI. Postmortem sperm procurement: a legal perspective. *Journal of Urology* 1999;161:1840–3.
26. Ethics Committee of the American Society of Reproductive Medicine. Ethical considerations of assisted reproductive technologies. *Fertility and Sterility* 1997;67(suppl): 1–9S.

16

Education

A Choice Between Public and Private Schools: What Next for School Vouchers?

Paul E. Peterson

For many years, fears that school vouchers were unconstitutional slowed their adoption by many state legislatures. But in 2002, the Supreme Court found in the case of Zelman v. Simmons-Harris that the Cleveland school voucher program was constitutional. The Court declared that the program did not violate the Establishment of Religion Clause of the U.S. Constitution, as plaintiffs had argued, because it allowed parents a choice among both religious and secular schools. There was no discrimination either in favor of or against religion. Now that school vouchers have passed this crucial constitutional test, many state legislators and other state officials are giving more thought to the voucher concept. In addition to Cleveland, experiments are underway in Milwaukee, Florida and Colorado and ate under active consideration in many other states. This essay seeks to answer some of the questions that are frequently raised.

SCHOOL VOUCHERS: THE BASICS

Simply defined, a voucher is a coupon for the purchase of a particular good or service. Unlike a ten dollar bill, it cannot be used for any purpose whatsoever. Its use is limited to the terms designated by the voucher. But like a ten dollar bill, vouchers typically offer recipients a choice. For this reason, distant relatives find coupons popular birthday presents for those whose tastes are unknown. The birthday child can be given a toy store coupon, without dictating the exact game or puzzle.

It is not only in the business world that vouchers or coupons are used. Food stamps, housing allowances for the poor and federal grants for needy students are all voucher-like programs that fund services while giving recipients a range of choice. Now, the idea is being advanced as a way of enhancing school choice as well. If parents are given a school voucher, the money will certainly be spent on education. But instead of requiring attendance at the neighborhood school, no matter how deficient, the family is given a choice among public and private schools in its community.

In other words, a school voucher is something like a scholarship to be used at one's choice of school. Indeed, in the United States there are numerous privately funded scholarship programs that operate much like school voucher programs. They allow the parent to pick the private school of their choice, but they pay approximately half the tuition for more than 60,000 students in New York City, Washington, D.C., Dayton, Ohio and many other cities across the country.

Although these private programs have generated valuable information about school vouchers, as discussed later in this essay, more important are the publicly funded ones enrolling over 25,000 students in Milwaukee, Cleveland and Florida. Colorado's newly enacted voucher program is to begin in the fall of 2004. All of the programs are restricted to low-income or otherwise disadvantaged children.

The oldest program, established in Milwaukee in 1990 at the urging of local black leaders and Gov. Tommy Thompson, was originally restricted to secular private schools and to fewer than 1,000 students. In 1998, the Wisconsin Supreme Court ruled constitutional a much larger program that allowed students to attend religious schools as well. In 2002–03, over 11,000 students, more than 15 percent of the eligible population, were receiving vouchers up to $5,783, making it the country's largest and most firmly established voucher program.

The Cleveland program, enacted in 1996, was of lesser significance until the Supreme Court made it famous. Before the decision ruling it constitutional, vouchers amounted to no more than $2,250 and were limited to approximately 4,000 students. After the Supreme Court decision, the number of students increased to over 5,000 and the amount of the voucher in fall 2003 could go as high as $2,700.

The initial Florida program, established in 1999 after Gov. Jeb Bush had campaigned on the issue, initially had less than 100 students but is poised to become somewhat larger. In this program, vouchers are offered to low-income students attending failing public schools. (The Colorado program, scheduled to go into effect in 2004, has a similar focus.) Initially, only two schools in Pensacola were said to be failing, but in 2002, 10 more joined their ranks. A second Florida program, which offers vouchers to students eligible for special education services, has received less attention but is perhaps more significant. In 2002–03, over 8,000 of Florida's special education students were enrolled in nearly 500 private schools.

In other words, a variety of privately and publicly funded voucher programs are in operation. Much can be learned from taking a closer look at how they operate in practice.

A FOCUS ON LOW-INCOME, MINORITY FAMILIES

Most voucher programs are focused on low-income or otherwise disadvantaged families, because their children are the ones least well served by traditional public schools. Voucher proponents point out that middle-income whites can pick their school by moving into a desired neighborhood or using a private school, while low-income blacks cannot easily do so. As voucher proponents love to point out, school choice is already part and parcel of the American educational system. Every time parents identify a neighborhood to live in, they select a school for their child—often self-consciously. According to a recent survey, 45 percent of whites (as compared with 22 percent of African-Americans) consider "the quality of the public schools" when deciding where to live.

Since African-Americans have the least amount of choice among public schools, they benefit the most when choice is expanded. In evaluations of private voucher programs in New York City, Washington, D.C. and Dayton, Ohio, my colleagues and I found that African-American students, when given a choice of private school, scored significantly higher on standardized tests than comparable students remaining in public school. In New York, where estimates are most precise, African-American students who switched from public to private schools tested, after three years, roughly 8 percentage points higher than African-Americans in public schools—nearly a two grade level improvement. These test score gains were accomplished at religious and other private schools that had much less money than that available to New York's public schools. Data available from the state of New York reveals that New York City's public schools have twice as much money per pupil as Catholic schools do—even after deducting amounts spent on the food lunch program, special education, transportation-related expenditures and the cost of the city's massive public school bureaucracy. With so little money, these schools do not have fancy buildings and playgrounds. Indeed, private school parents reported fewer facilities and programs at their child's school than public school parents did.

Yet private school parents also reported much higher levels of school satisfaction than their public school peers. Private school parents also were more likely to report that their child had smaller schools, smaller classes and an education-friendly environment (less fighting, cheating, property destruction, truancy, tardiness and racial conflict). Their children had more homework and the schools were more likely to communicate with the family. Nor were the private schools any more segregated than the public ones. There was no evidence that vouchers improved the test scores of students from other ethnic groups, however. Vouchers did not have a significant impact, positive or negative, on the test scores of either whites in Dayton of Latinos in New York City.

These findings are all the more important, because they come from randomized field trials similar to the pill-placebo trials conducted in medical research, generally regarded as the gold standard of scientific research. Yet the results from these randomized field trials do not so much break new ground as confirm findings from other studies. In a review of the broad range of research, Jeffrey Grogger and Derek Neal, economists from the University of Wisconsin and University of Chicago, find that "urban minorities in Catholic schools fare much better than similar students in public schools," but the effects for urban whites and suburban students generally are "at best mixed."

It is little wonder that many African-Americans are among those most eager to find alternatives to traditional neighborhood public schools. Even though many civil rights leaders oppose school choice, a majority of their constituents think otherwise. In 2000, the Joint Center for Political and Economic Studies reported that 57 percent of African-American adults favored vouchers, as compared with 49 percent of the overall population.

NO CHILD LEFT BEHIND

If students who attend private schools seem to benefit thereby, how about those students left behind in traditional public schools? To answer this question, one needs to consider the students in the voucher program, the academic impact on public schools and the financial impact on public schools.

Do Vouchers Attract the Best and the Brightest?

My own research has looked at this question in two different ways. In one study, my colleagues and I compared a cross-section of all those who applied for a voucher offered nationwide by the Children's Scholarship Fund with a comparable group of those eligible to apply. African-American students were twice as likely to apply as others. Specifically, 49 percent of the applicants were African-American, even though they constituted just 26 percent of the eligible population. Other results reveal little sign that the interest in vouchers is limited to only the most talented. On the contrary, voucher applicants were just as likely to have a child who had a learning disability as non-applicants. Additionally, participants were only slightly better educated than non-applicants.

In New York, Washington, D.C. and Dayton, my colleagues and I found no evidence that private schools' admission policies discriminated on the basis of a young student's test score performance. Only among older students (grades 6–8) in Washington, D.C., did we see some signs that private schools expected students to meet a minimum educational standard prior to admission.

Other researchers find much the same pattern. In Milwaukee, the Wisconsin Legislative Audit Bureau found that the ethnic composition of the participants in Milwaukee's voucher program during the 1998–99 school year did not differ materially from that of students remaining in public schools. Also in Cleveland, Indiana University analysts said that voucher "students, like their families, are very similar to their public school counterparts."

Upon reflection, these findings are not particularly surprising. Families are more likely to want to opt out of a school if their child is doing badly than if that child is doing well. A number of families, moreover, select a private school because they like the religious education it provides, or because it is safe, or because they like the discipline. When all these factors operate simultaneously, the type of student who takes a voucher usually does not look different from those who pass up the opportunity, except perhaps for the fact that those within a specific religious tradition are more likely to choose schools of their own faith.

Public School Performance

If vouchers do not simply pick off the top students within the public schools, but attract instead a broad range of students, then there is no obvious educational reason why public schools should suffer as a result of the initiative. On the contrary, public schools, confronted by the possibility that they could lose substantial numbers of students to competing schools within the community, might well pull up their socks and reach out more effectively to those they are serving. Interestingly enough, there is already some evidence that public schools do exactly that.

Harvard economist Caroline Minter Hoxby has shown, for example, that since the Milwaukee voucher program was established on a larger scale in 1998, it has had a positive impact on public school test scores. The public schools in the low-income neighborhoods most intensely impacted by the voucher program increased their performance by a larger amount than scores in areas of Milwaukee and elsewhere in Wisconsin not affected by the voucher program.

Even the threat of a voucher can have a positive effect on test scores. Research by Manhattan Institute scholar Jay Greene shows that when public schools were in danger of failing twice on the statewide Florida exam, making their students eligible for vouchers, these public schools made special efforts to avoid failure.

Fiscal Impacts on Public School Children

To see how school vouchers affect the fiscal resources available to public school children, the structure of public school financing needs to be briefly considered. Although the financial arrangements vary from one state to the next, on average, nationwide, 49 percent of the revenue for public elementary and secondary schools comes from state governments, while 44 percent is collected from local sources, the balance received in grants from the federal government. Most of the revenue school districts get from state governments is distributed on a "follow the child" principle. The more students in a district, the more money it receives from the state. If a child moves to another district, the state money follows the child. Local revenue, most of which comes from the local property tax, stays at home, no matter where the child goes. As a result, the amount of money the district has per pupil actually increases if a district suffers a net loss of students, simply because local revenues can now be spread over fewer pupils.

The voucher programs in Milwaukee, Cleveland and Florida have been designed along similar lines. The state money follows the child, but the local revenue stays behind in local public schools, which means that more money is available per pupil. In Milwaukee, per pupil expenditures for public school children increased by 22 percent between 1990 and 1999, rising from $7,559 to $9,036. Not all of the increase was a direct result of the voucher program, but the example shows that public schools do not necessarily suffer financially when voucher programs ate put into effect.

Balkanization: Myth, Not Reality

Whatever the advantages of vouchers, some may feel that they would prove divisive in a pluralist society with multiple religious traditions. In his dissent from the majority opinion in Zelman, Justice Stephen Breyer saw the decision as risking a "struggle of sect against sect." And Justice John Stevens said he had reached his decision by reflecting on the "decisions of neighbors in the Balkans, Northern Ireland, and the Middle East to mistrust one another. . . . [With this decision] we increase the risk of religious strife and weaken the foundation of our democracy."

These dissents echo the concerns of many distressed by the worldwide rise in fundamentalist religious conviction, worries that have intensified since 9/11. But though the concerns are genuine enough, it's hardly clear that government-controlled indoctrination of young people is the best tool for conquering intolerance. On the contrary, this strategy proved counterproductive in many parts of the former Soviet Union. Historically, the United States has achieved religious peace not by imposing a common culture but by ensuring that all creeds, even those judged as dangerous by the enlightened, have equal access to democratic processes.

Of course, religious conflict is part and parcel of American political history. In the late 19th century, many objected to the establishment of Catholic schools. Indeed, anti-immigrant sentiment was so strong that amendments to state constitutions were enacted that seemed to forbid aid to religious schools. Many of these provisions are so-called "Blaine" amendments, dating to the 19th century, when James Blaine, a senator from Maine and a Republican presidential candidate, sought to win the anti-immigrant vote by campaigning to deny public funds to Catholic schools. (Blaine is perhaps most famous for tolerating a description of Democrats as the party of "Rum, Romanism, and Rebellion.") In its classic version, the Blaine amendment read as follows: No money raised by taxation for the support of public schools, or derived from any public fund therefore, nor any public lands devoted thereto, shall ever be under the control of any religious sect; nor shall any money so raised or lands so devoted be divided between religious sects of denominations.

Blaine-like clauses in state constitutions are being invoked by those seeking to forestall voucher initiatives. In a number of cases, state courts have interpreted these clauses to mean nothing more than what the Supreme Court defines as the meaning of the establishment clause of the First Amendment. If this view prevails in state courts, then vouchers do not violate these state constitutional clauses now that they have been found constitutional by the U.S. Supreme Court. But not every state judge necessarily shares this view. Such language has proven to be a hurdle for the voucher program in Florida, for example, where a trial court has found the law in violation of the state constitution. Depending on what happens to the appeal of this trial court decision, the U.S. Supreme Court may eventually be asked to decide whether, on account of their nativist and anti-Catholic origins, the Blaine amendments—and their derivatives—are themselves unconstitutional.

The controversies over religion seem more heated in the political and legal world than in the classroom, however. While exceptional cases can always be identified; there is little evidence that religious schools typically teach intolerance. Indeed, careful studies have shown that students educated in Catholic schools are both more engaged in political and community life and more tolerant of others than public school students. After enduring harsh criticism from critics in a Protestant-dominated America, Catholic schools took special pains to teach democratic values. The more recently established Christian, Orthodox Jewish and Muslim schools can be expected to make similar attempts to prove they too, can create good citizens.

As Justice Sandra Day O'Connor pointed out in her concurring opinion, if Breyer and Stevens' fears were real, we would know it already. She showed that taxpayer dollars flow to religious institutions in multiple ways—through Pell Grants to sectarian colleges and universities; via child care programs in which churches, synagogues and other religious institutions may participate; and through direct aid to parochial schools of computers and other instructional materials. If thriving religious institutions create a Balkanized country, she seems to say, this would already have happened.

Nor, say voucher proponents, have public schools eliminated social divisions. As Clarence Thomas argued in his concurring opinion, "The failure to provide education to poor urban children perpetuates a vicious cycle of poverty, dependence, criminality and alienation that continues for the remainder of their lives. If society cannot end racial discrimination, at least it can arm minorities with the education to defend themselves from some of discrimination's effects." In other words, vouchers may help heal, not intensify, the country's most serious social division.

17

Religion

And the Wisdom to Know the Difference? Freedom, Control and the Sociology of Religion

Eileen Barker

This is a theoretical essay, but it is grounded in the empirical observations of the sociology of religion; and it has a political twist to it. The main thrust of the argument rests on the assumption that freedom is a socially relative concept, and, as such, not only can it be both increased and decreased according to circumstances, it can also be increased or decreased through a knowledge or ignorance of such circumstances. Questions are asked about the ability of individuals to choose their own religion, some of the potential consequences of their choices, and ways in which the state apparatus can regulate religious groups. The not-so-hidden agenda is that social scientists might further contribute to our freedom through an increased understanding of those things that we cannot change, and of how best to change those things we can—and, perhaps, of our wisdom to know the difference.

> God grant me the Serenity to accept the things I cannot change, the Courage to change those things I can, and the Wisdom to know the difference.
>
> (Reinhold Niebuhr 1892–1971)

The sociology of religion has long been concerned with issues of freedom and control. At an ontological and somewhat speculative level, there have been numerous treatises on whether it is the lot of individual participants in a society to be active actors ruling their own destinies or merely passive recipients, subject to the vagaries of social forces. More empirically, the concern has been evidenced by work at individual, group and societal levels on, respectively, subjects such as brainwashing, mind control and/or mental manipulation; types of authority wielded by religious leaders and institutions; and/or the regulation and control of religions by states. Some of us have been particularly concerned with recent developments in post-communist countries, and that is the subject on which I chose to concentrate when selecting the contributors for this special edition of the journal.

We have, however, still a long way to go in understanding the processes that broaden and narrow our ability to determine (in both senses) the patterns of our lives. With one or two notable exceptions, our work has tended to focus on static situations and has been restricted to, a series of particular, discrete circumstances. We have, for example, confined our findings to statements such as: In situation A (when, say, a small religious group is in a closed environment, cut off from the rest of society) freedom and/or control has tended to be advanced or curtailed according to some unspecified, non-comparative scale.

Or: In situation B (when, say, there is a strong relationship between the state and one traditional religion) the freedom and/or control of co-existing minority religions has tended to be diminished according to another unspecified and non-comparative scale.

Such conclusions do, of course, provide us with important and useful information. There does, none the less, seem to be a paucity of empirical studies that embrace a wider frame of reference. That is, there is relatively little work that, systematically,

(a) compares types and degrees of freedoms between, say, the Amish, Roman Catholics, Buddhists and Unificationists—let alone the Amish in nineteenth century Canada; the Catholics in fifteenth century Italy; Buddhists in seventeenth century Thailand, and Unificationists in twenty-first century Japan.

(b) examines the dynamics of a process in which A moves to B and then progresses to C—when, for example, freedom is decreased through an increase in the number and application of regulations, but this results in a reaction that overthrows the regulating authority, thus resulting in (perhaps) greater freedom.

(c) explores the complexities of situations in which one person's or group's liberty depends on curbing the liberty of other persons or groups.

(d) pursues the empirical relationships involved in the philosophically familiar distinction between "freedom from" and "freedom to."

THE PARAMETERS OF FREEDOM

Fatalism, Brainwashing and Total Freedom

There are numerous ways in which we can conceive of, and ask questions about, freedom and control. These are, of course, concepts that lie at the very heart of the sociological enterprise. Marx, Durkheim, Weber, Simmel and Mead might have started from different assumptions—and they certainly came up with different answers—but they were all concerned with the ways in which individuals and groups are, variously, enabled and restrained by the structures and cultures within which they find themselves, and how they create, conserve, change and negotiate those structures and cultures.

For even longer, the concepts of freedom and determinism have lain at the root of the philosophical enterprise. They have been variously associated with further concepts such as causation and control; chance and choice; chaos and confusion, the first of each pair being usually, but not necessarily, taken to apply to an objective state, while the second refers more to subjective states involving human agency. Sometimes Heisenberg's Uncertainty Principle is invoked to explain how we can have choice: as not everything is determined, the argument goes, we are free to make decisions. In fact, if there were no regularities beyond the sub-atomic levels of organisation that Heisenberg was talking about, there would be nothing but chance occurrences,[1] leading to chaos.[2] Our ability to control our lives depends on our perceiving causes (that X leads to y);[3] if there were no patterns of behaviour at the social level, there would be confusion rather than choice.

One rather irritating perspective is the unadulterated tautology of fatalism, which, a priori, rules out the possibility of freedom. Everything we do, the argument goes, is determined in one way or another—as is our belief that we are free to make our own choices.[4] This may, of course, be true. But if it is, it is also true that most of us are "determined" to muse on and even to investigate what we consider to be something that it makes some sort of sense to call freedom.

Approaching the question from a very different philosophical perspective, that of ethics and moral philosophy, we meet a contrary position. The very idea that it makes sense to talk about morality (about performing a right action rather than a wrong action and/or being a good person rather than a bad person) presupposes that an individual not only has the option of choosing between at least two alternatives, but also that s/he is responsible for that choice. In other words, the very concept of morality assumes that toe are free to restrict our freedoms ourselves; we consider that we and others ought not to perform certain actions that it is perfectly possible for us to perform. To put the argument slightly differently, some kind of criterion is needed to distinguish moral from immoral behaviour ("oughts" from "ought nots") within the realm of those potential actions that have not already been ruled out of court by other constraints—and as morality involves a denial of the legitimacy of some of those acts that we could perform, it follows that a moral decision involves a reduction of our freedom of action—and this is a decision that often rests on criteria supplied by religion.

Sometimes metaphors are used to portray different images of freedom and control. One that I quite like is to see ourselves as walking along a valley which has mountains on either side, and which becomes bifurcated by other mountains, thereby creating alternative valleys along which we might proceed. Sometimes the valleys are wide, and sometimes they are narrow; sometimes the mountains are steep and impenetrable, and sometimes they have gentle slopes; sometimes we can climb and expand the contours of the path, even climb to the top and look into the next valley; but eventually we encounter mountains that we cannot climb. There are always limits to our freedom—sometimes these limits are more and sometimes they are less than we are aware of.

Brainwashing is another kind of metaphor, but one that is frequently drawn upon not as an image but as an explanation. It rests on an assumption that the beliefs and/or behaviour of certain individuals are the result of their having had their freedom removed.[5] One category of individuals still popularly considered to be the victims of brainwashing is that of converts to the so-called "cults" who, it is claimed, have been turned into mindless robots, unable to think or make any choices for themselves. However, during thirty-plus years of studying new religions and talking to thousands, if not tens of thousands, of the movements' members, I have yet to meet a fully automated robot. I have, rather, become convinced that the overwhelming majority of those who resort to the metaphor of brainwashing are more likely to be motivated by their incomprehension or downright distaste of the content of the alternative belief system than by their comprehension of the actual process of conversion to it.

This conviction arises from research I began in the mid 1970s, when accusations of brainwashing were at their height and being used to justify deprogramming—the illegal practice of kidnapping members of new religious movements (NRMs) and holding them against their will until they escaped or renounced their faith. My interest in the subject was motivated as a citizen concerned about freedom as well as by the challenge presented to me as a sociologist concerned about the relationship between the individual and the social group. What I was initially unsure about was whether a restriction of the individuals' freedom would be greater as a consequence of a process of brainwashing or of deprogramming. If the Unificationists were really using brainwashing or mind control techniques, then surely this ought to be properly investigated so that something could be done to prevent such a blatant removal of people's freedom of choice—there might then even be an argument for "deprogramming." But if this were the case, then the practice should be conducted by properly trained people rather than those who were at the time violently depriving the converts of their freedom in the name of liberating them—often with harmful consequences.[6]

Faced with the challenge of going beyond circular a priori assumptions that converts (or "recruits") were either free or coerced when they joined a movement such as the Reverend Sun Myung Moon's Unification Church,[7] I decided that, while distinguishing between freedom and determinism involved two concepts that appeared well nigh impossible to operationalise, it might be possible to operationalise the concept of choice, at least for the purposes of this particular study insofar as it focussed on processes happening in a clearly defined situation over a limited period of time.

> A choice would involve reflection (in the present), memory (of the past) and imagination (of possible futures). A person would be an active agent in deciding between two or more possible options when [s]he could anticipate their potential existence, and when, in doing so, [s]he drew on his[her] previous experience and his[her] previously formed values and interests to guide his[her] judgement (Barker 1984:137).

The social situation in question arose when potential members, who had already become persuaded that the Unification Church might have something worthwhile to offer them, were invited by the movement to attend a residential workshop. Here they would be subjected to the so-called mind control techniques. The proposition (promoted by certain sections of the media and "anticultists") to be tested was that irresistible and irreversible practices employed during the workshop would result in all those who had accepted the invitation ending up as compliant "Moonies," ready to devote their lives to the every whim of their Korean Messiah, Sun Myung Moon.

What transpired was that, of over a thousand who had attended workshops in the London area during. 1979, 90 percent did not end up joining the movement. It seemed clear, therefore, that I had to look at the individuals (rather than merely at the techniques employed during the workshops) if I were to understand why some would join and others would not. In other words, the obvious enough conclusion was that the Unification environment (its techniques and its promises) could not by itself provide a sufficient explanation for the outcome; it was necessary to take the individuals themselves into account as independent variables in the process.[8]

It may well be the case that Unificationists and committed members of many other new religions would like to be able to have more success in drawing new members to their faith. It is certainly not being denied that they tried hard to influence those who were prepared to listen, and that they could be both persistent and aggressive (and sometimes deceptive) in their proselytising efforts. Scholarly research has, however, shown over and over again that the techniques of persuasion employed by the NRMs are a far less efficient means of gaining new members than the practice of being born into an established tradition, such as Islam or Roman Catholicism. The fact is, however, that (a) the vast majority of people subjected to "brainwashing" find themselves perfectly capable of saying "no, thank you," and that (b) those who do join the NRMs (and, presumably, are subjected to even greater doses of such techniques) have managed to leave of their own free will (Bromley 1988; Wright 1987). This must suggest, at least in situations which involve no physical restrictions or malfunctioning of the brain, that there is, minimally, some collaboration on the part of the individual.

Resting on an assumption diametrically opposed to that underlying the metaphor of brainwashing is the belief we can achieve total freedom. This is what the Church of Scientology offers its practitioners, the proposition being that we can, through the techniques of Dianetics and pursuing the truths revealed by L. Ron Hubbard, reach such a state (Wallis 1976).

> Scientology offers humanity freedom from this needless suffering, both now and for all future time. By following the path outlined in the scripture of the Scientology religion, the thetan[9] can progress through higher and higher levels of spiritual awareness and return to his native state and thereby achieve complete spiritual freedom. Now, in this lifetime complete spiritual freedom can be achieved. The primary path to this spiritual freedom is through "auditing," one of the two central religious practices of the Scientology theology. With this freedom comes release from the eternal cycle of birth and death and full awareness, memory and ability independent of the flesh. And with it comes a spiritual being who is "knowing and willing cause [sic] over life, thought, matter, energy, space and time." (Hubbard 1998:562)

But while Scientology claims that it differs from the Jewish and Christian traditions in that its concept of salvation is much more immediate, those of us who have yet to encounter a "fully operating thetan" may suspect, as doubting Thomases, that (at least epistemologically) the offer of total freedom in this life is not all that different from the offer of salvation in the next one. Neither seems likely to be put to the test in the immediate future. For that reason, at least until we are proved wrong, it seems more helpful to take the pragmatic approach of considering freedom as something that will be more or less present—rather than totally present or totally absent. And this now takes us back to the question: What reduces and what increases freedom?

Laws, Regularities and Degrees of Modifiability

It has often been pointed out that knowledge of the ways in which we are not free can itself give us a kind of freedom—the serenity to accept the things we cannot change—or, more crudely, the freedom not to bang our heads against a brick wall. The natural sciences describe laws that clearly impose well nigh insuperable limits on our freedom. We are bound by our human bodies; we cannot escape the passage of time; we cannot deny the pull of gravity; we cannot live without food and water. Freedom from such restrictions would take us beyond this world as most of us know it—although some religions such as Shamanism teach this is not impossible, and even non-believers have reported out of body experiences.

So far as the laws of nature are concerned, however, there is general agreement in normal scientific discourse that we cannot change them. Their existence is independent of our existence. None the less, through our knowledge and understanding of these laws, we can find a way around them; it is through our use of the laws that we can overcome some of the constraints that they have imposed on us and give ourselves new freedoms. It is insofar as we understand the functioning of gravity that we can escape the earth's atmosphere and get to the moon; it is insofar as we understand how cancer "works" that we can hope to cure the disease.

But as we move from the apparently immutable laws of physics, into chemistry, biology and then the human sciences, we enter areas of ever-increasing complexity and modifiability. As further levels of organization evolve, new properties emerge—by which I mean new constraints and new potentialities; new controls and new freedoms. Returning to my earlier metaphor, as we move within the ever-increasing complexities of social life, each level of organization forms yet another range of mountains, which, although in some ways more negotiable than the previous range, has, nevertheless, still the ability to restrict our potential actions in ways they would not otherwise have been restricted. Thus, there is nothing in the laws of physics that would prevent our driving on the "wrong" side of the road, or strangling our children, yet most of us are prevented by moral codes or the law of the land from performing such actions. But these are different kinds of restraint from the pull of gravity—and some people do drive on the wrong side of the road, and some people do strangle their children.

It is insofar as people participate in and share a socially constructed reality that we can detect regularities in the social sciences (Berger and Luckmann 1967).[10] Social life itself would not be possible if there were not some tacit agreements about what the world is like; and how things ought to be done. These more or less shared perceptions of how things are and how they ought to be give rise to regularities in our behaviour, and allow us to predict how others will behave—but only up to a point, and usually as a statistical probability rather than the absolute, universal, all-time, all-places certainty that is, to most intents and purposes, the case with the natural sciences. In other words, the precise nature of the reality we construct, share, negotiate and change differs (to a greater or lesser degree) from time to time, from individual to individual, from group to group, from society to society—and from religion to religion. And this means that opportunities for freedom of individual choice differ from situation to situation.

Just as we could use our knowledge of the laws of nature to vary their operation, there is a potential for manipulation of social life through an understanding of how society operates. But there is also a difference. Not only can we use our knowledge of "social laws"—or, more warily, regular relationships between variables—to control outcomes, we can also change the regularities themselves so that they no longer pertain. This is partly because the regularities that occur in social life are, obviously enough, dependent for their existence upon our existence. When talking about social regularities, as opposed to the laws of physics, we might be wise to move from the ontological statement: "if X changes in a particular way, then Y will change in a particular way" to the more cautious epistemological statement: "if X changes in a particular way, then Y is likely to change in a more or less predictable way." We can be confident that water boiled at 100 degrees centigrade under conditions of normal pressure in Jerusalem in the fifth century CE, just as it did in nineteenth century Chicago. It is likely that the price of fish increased when the supply fell in both societies, but a number of other variables might have intervened: a dramatic fall in demand might have been due to the discovery that the local fish were poisoned, or to the introduction of a religious doctrine that believers should adopt a vegetarian diet. Such factors affecting demand would make the correlation between price and supply far more unstable and subject to the ceteris paribus clause. It is because social reality is an on-going process of construction, mediated through individuals with different experiences, perceptions, and interests, that it is relative to time and place—and new "knowledge" can lead to changes in regularities, which can both enhance and diminish the freedom of both individuals and groups (Popper 1961:vi).

RELIGION AND FREEDOM

Religion can both expand and foreclose our horizons. It can enable us to glimpse the otherwise unimaginable—to dream the impossible dream. But it can also banish dreams from our minds by its strict and restrictive injunctions. Throughout history it has been one of the primary agents for sorting out the confusions of the human predicament by imposing order upon our understanding of what the world is like and what it should be like. But this means that it has also had a significant control over its members, restricting their freedoms through invoking feelings of guilt, or by burning them at the stake. Religious values have inspired men and women such as Albert Schweitzer and Mother Theresa. Religious edicts have underpinned suicide bombings, amputations, female infanticide and genital mutilation, and the practice of suttee. Promises of heavenly paradise in the hereafter have been juxtaposed with threats of hell fire and eternal damnation—or the prospect of being reborn as a woman.

Of course, social scientists must remain methodologically agnostic; they cannot draw on God or any other super-natural phenomenon as an independent variable. But if they want to understand ways in which freedoms can be enhanced and curtailed they need to understand the beliefs that motivate and dampen actions. Sometimes a strong religious belief leads directly to accepting, without question, a restriction of choice. In such cases, the belief can be that the decision has been made by God, or the gods, and no alternative (and, thus, no freedom of choice) is open to the believer. At one extreme there is the Calvinist doctrine of predestination (which, Weber (1930) argued, led to the rise of modern capitalism and all the new freedoms and constraints that it brought in its wake). Somewhat less immutable than predestination is the doctrine of karma, which explains why we find ourselves in the situation into which we were born—but in some sense the choice was outs, even though we may have been unaware at the time that we were deciding our destiny. We can, however, become aware of what the consequences of our actions will be for the future and, through following the correct path, make more enlightened choices.

Another position is that God's expectations are irresistible despite the appearance of choice—this was the gist of Luther's "ich kann nicht anders."[11] At the same time, there lies at the base of Christian theodicy the belief that God had given Adam and Eve freedom to sin in the Garden of Eden—and sin they not only could, but did. Then there is the belief that we no longer enjoy the freedom that is our birthright. What, it can be asked, are we to do in the face of the belief "L'homme est ne fibre, et partout il est dam les lets"?[12] Scientologists teach that the thetan is born free, but, because of the accumulation of "engrams" (blockages) over a number of lives, needs to become "Clear"—which is possible through the technology of Dianetics (Hubbard 1998:xx, 65ff). Yet another conviction is that the only real choice open to the believer is to submit to the God in "whose service is perfect freedom."[13] Be it for good (Durkheim 1968) or ill (Marx 1963), religious beliefs can also control our aspirations by adding an extra legitimacy to "the way things are" in society and, thereby, making us less likely to question the status quo. As the popular hymn, *All Things Bright and Beautiful,* puts it: The rich man in his castle, The poor man at his gate, God made them, high or lowly, And order'd their estate (Alexander 1848).

Pluralism and Individual Conversion

A frequent rejoinder to those who complain that belonging to a religion diminishes one's freedom is to say that one has freely chosen to accept religious membership and the restrictions associated with that religion. The subject of brainwashing removing that initial choice has already been discussed, but it should be noted that it is only relatively recently that, practically (or socially) speaking, the choice of religion was open to more than a few individuals—usually those in privileged positions. For the majority of the population, one's religion was determined by one's social superiors. Thus, if the monarch changed his or her religion (because, to take the case of Henry VIII, he wanted to divorce Catherine of Aragon in order to marry Anne Boleyn), then his subjects were expected to change too. A new king, queen or lord as a result of conquest (as at the time of the Crusades) or inheritance (as with the ascension of Mary Tudor to the throne of England) could also lead to an overnight change in an entire population's religious allegiance. Indeed, an individual could be "converted" several times in his or her life on account of the institution of cuius regio eius religio.[14] Shifting our language once again, the concept of Une foi, an loi, un roi (one faith, one law, one king) epitomises the way in which the state, society, and religion were all bound up together in people's minds and experience. There was not the distinction that we now take for granted between public and private, or between civic and personal:

> "One faith" was viewed as essential to civil order—how else would society hold together? And without the right faith, pleasing to God who upholds the natural order, there was sure to be disaster. Heresy was treason, and vice versa. Religious toleration, which to us seems such a necessary virtue in public life, was considered tantamount to letting drug dealers move next door and corrupt your children, a view for the cynical and world-weary who had forgotten God and no longer cared about the health of society (Le Poulet Gauche 1998:5/2).

This is not to say that individuals of lowly status might not convert—many of the new religions of yesteryear appealed disproportionately to the oppressed classes (Cohn 1970; Worsley 1968)—but there was often a high price to pay. To choose to convert might involve the "choice" of being thrown to the lions for an early Christian, or being burned at the stake for a medieval Cathar. Less dramatically, those who chose not to attend church in the reign of Elizabeth I would be freed one shilling—a sum that could be accepted as a religious tax for aristocratic Catholics, but was hardly a viable option for lowly peasants.

It can be argued that it was the nineteenth century Protestant missionaries from Europe and North America who were first concerned with the widespread conversion of individual souls, rather than a mass switch motivated from the top.[15] At the same time, the ideology of individualism, freedom and choice had been becoming increasingly prevalent.[16] On the one hand, with the growth of modernization, personal identity has become increasingly associated with individual achievements and, on the other hand, authority figures of yore, including religious leaders, have been commanding less respect. This can be associated with, among other things, changes in family life, educational methods (with a shifting focus from rote learning to personal enquiry and experimentation), and the phenomenal growth of the mass media which has both undermined the authority of political and religious leaders and offered a previously inaccessible wealth of alternative ideas to anyone who can read or access a radio transmitter, a television—or, now, the world wide web. The significance of these changes should not be exaggerated, but they certainly cannot be ignored. There can be little doubt that the concept of individual freedom has reached an unprecedented pervasiveness with the rhetoric and practice of hitherto unimaginable proportions. But the freedom with which contemporary Western individuals and groups are faced can be a double-edged sword. As already intimated, freedom is a multifaceted concept embodying many complexities.

TENSIONS BETWEEN COMPETING FREEDOMS

In the contemporary quest for freedom it is possible to observe a seemingly contradictory relationship between freedom on the one hand, and choice and control on the other. The contradiction lies in the fact that either too much or too little choice and/or control can lead to a diminution of freedom. In other words, while it might seem fairly obvious that removing choices and imposing controls can decrease freedom, and that increasing choices and reducing control can increase freedom, it is also true that reducing choices and imposing controls can result in a subjective experience of more rather than less freedom; and reducing controls and having more choices can result in less rather than more freedom.

The theoretical reasoning behind this statement starts from the definition I suggested earlier: (a) choice depends on our being able to anticipate possible futures; (b) any reliable anticipation must depend on order and predictability; (c) Just as one cannot have choice without order and predictability, one cannot have freedom without control.[17] Let us, however, look at some empirical examples.

It has been suggested that the contemporary West (Europe, North America, Australia and New Zealand) is one of those times and places when traditional religious authority has been relatively ineffective as a controlling influence in individuals' lives.[18] While the vast majority of people remain in the religion into which they were born, the West has seen an increasing number choosing to move away from traditional institutional religion in a number of directions. These may be divided into five main ideal types. First, there are what may be called the "hard secularists" who reject the existence of God or gods and who may belong to some Humanist Association or follow a Marxist/Leninist or other atheistic ideology. Secondly, there are the "soft secularists" who differ significantly from the previous type in that they do not deny the existence of God, but see no reason to attend a place of worship or bring religion into their lives—except on occasions of stress or, perhaps, as part of a rite de passage, at which times they expect the professionals to be there to provide whatever support is needed or appropriate. Thirdly, there are fundamentalist groups that have a strong belief in The Truth and who tend to separate themselves from the trot of society with a strong social, and sometimes geographical, boundary. Unlike the soft secularists, their religion is a primary source of their identity and has a bearing on pretty well every aspect of their highly structured lives. In these groups there is relatively little room for individuals to interpret their beliefs or negotiate their behaviour, such matters being decided by the group's scripture and/or leaders. The fourth type is in many ways diametrically opposed to the third. It is one in which, to use its own language, spirituality rather than religiosity is celebrated. God is perceived as something within each individual rather than "out there." External authority is rejected in favour of personal responsibility; bounded groups are replaced by networks of the like-minded; and there is frequently a stress on the feminine and ecological values. Fifthly, there are the NRMs, about which it is impossible to generalise, but which can be seen to span all the previous types, but particularly those of fundamentalism and spirituality.

A point that needs to be made here is that the diversity of religion in contemporary globalizing society reflects the diversity to be found in individual lives. Not only do people differ from others within their own societies, but their lives differ, often dramatically, from those of their parents. To paint two vastly over-simplified caricatures: some individuals experience their society as chaotic and/or libertarian, with no standards and no purpose; others experience the same society as an authoritarian, bureaucratic rat race within which we have to play roles that are imposed upon us by our parents, the educational system and "them," all of whom expect us to conform to restrictive norms and controls.

It is individuals of the first type who may feel that their freedom is increased as their choices are removed. They yearn to be informed that they are free to go one way only—The Way. In such an environment (which is on offer from new and old fundamentalist religions), one is unencumbered by the doubts and uncertainties of not knowing who one is, where one belongs, what to do, or how or why to do it. If one recognises the confusion that too much choice can bring in its wake, it is not all that difficult to understand the attraction of religions that are considered exploitative, oppressive or totalitarian by non-members, and it makes perfectly good sense to talk about choosing to have the freedom to be controlled. I have spoken to numerous members of the more authoritarian and controlling new religions (and many in more established fundamentalist groups) who have found it an enormous relief to be able to develop within an environment in which they do not constantly have to make decisions. This can be true whether the decision concerns the choice of a marriage partner or what toothpaste to use—whether one is talking about an ultimate goal (such as salvation) or the means to achieve that goal (by doing what the Messiah orders, by chanting a sacred mantra for so many hours a day, or by witnessing to others who need to be saved). To be free from the responsibility of making such decisions appears to afford large numbers of individuals the opportunity to become free to develop their lives in ways that would not have been available to them in "normal" circumstances.[19]

Like the representatives of the first caricature, those of the second type are also on a quest to be "free from" in order to be "free to." But while the former choose to retreat into the bounded, even womblike nature of highly prescriptive and ordered religions in order to lead meaningful lives (in what I have called elsewhere "the freedom of the cage"), the latter feel the need to pursue their freedom in a diametrically opposite direction—and may find themselves ensnared in "the cage of freedom" (Barker 1995). In the hope of achieving self-development or self-realization, these individuals are persuaded, particularly by the rhetoric of sections of the Human Potential movement, that it is only by getting rid of all social controls and constraints that their "true selves" can be released and flourish. By rejecting the structures, rules and roles that have been imposed since childhood, the liberated "real me" will, it is claimed, emerge. However, while it is true that freeing oneself of some social and cultural restrictions can undoubtedly have a liberating effect, it is also true that too much "liberation" can become counter-productive—at least from a sociological perspective that sees human beings as fundamentally (though not only) social beings. Freedom has been envisaged as the opportunity to do anything, but, as

Durkheim (1952) argued in *Suicide* and elsewhere, the removal of restraints can lead to a situation of confusion or anomie. The "liberated individual" can regress back to the "freedom" of a pre-socialised child. As imbued values are rejected, one can find oneself without any standards to act as a benchmark for assessing where one stands in relation to oneself or to the social worm out there. Chaos, confusion and a loss of self-identity may ensue, with, perhaps, an escalating dependency upon the group or guru offering the "freedom."

There are, of course, many other kinds of tensions and balances with which freedom can present us in the religious sphere. To some extent we have already been discussing choices between competing freedoms (a dilemma dissolved by those who believe in total freedom or total absence of freedom). At its most stark we are faced with the choice between this freedom and that freedom: if we wish to lead a life of devotion in an enclosed monastery, for example, we might have to give up several freedoms we would otherwise enjoy. More usually we have to prioritise our choices, and it can be argued that we have greater overall freedom insofar as we allow ourselves to be inconsistent in our choices, rather than sticking rigidly to the same hierarchy in every situation.[20]

Our Freedoms and Their Freedoms

Moving from the individual to the group level, one of the fiercest tensions throughout the history of religion has been that between our freedom and their freedom. This can take a number of forms, some of which do not involve violence, but strong and genuine beliefs that one person or group knows better than another group what is in the best interest of that other, the former forcing the latter to make a choice—but the option of democracy may not be what a theocratically minded nation wants.

Another example concerns the freedom of parents to bring up their children according to their beliefs. This can give rise to the potentially opposing freedom of children to make up their own minds. It has also led well-intentioned courts, acting they believe in the best interests of the child, to grant custody to one parent, rather than the other, who belongs to an unpopular religion but who, according to a number of other generally accepted criteria, would appear to be the more appropriate parent to bring up the child. There have also been cases (in Australia, Canada and elsewhere) when children have been removed by the state from indigenous groups to residential schools and/or adopted for rearing by "more civilized" guardians. Still on the subject of children, there is the dilemma faced by some liberally minded parents of whether taking their children to a place of worship on a regular basis gives them more or less choice to accept or reject that (or any other) religion in the future than not taking them to the services would.

Brief mention has already been made of how, when the Berlin Wall came down, the concept of freedom was on everyone's lips, and that of freedom of religion was one of the more frequently heard catch phrases in the rhetoric of the time. The countries of Eastern Europe and the Former Soviet Union had been living under atheistic regimes that had, to a greater or lesser extent, imposed severe restrictions on religious institutions and practices. Priests had been imprisoned and murdered; churches and mosques destroyed or used for secular purposes; those who were known to be religious were denied advancement in their careers and their children were denied university education. Once the state-imposed secularism was removed, the traditional religions not unnaturally wanted to reclaim their property—and their flock.

They faced a great number of problems, however. These included poverty, lack of experience in pastoral and educational skills; an ageing priesthood; accusations of collaboration and a theologically ignorant population. A further problem that soon became apparent was the massive influx of proselytising religions from the West which had scrambled over the rubble of the Berlin Wall to offer their religious and spiritual wares to fill what they perceived to be a gaping ideological vacuum. Some of these religions had been operating underground during the Soviet period, but now the new missionaries were, the Mother Churches complained, bribing their innocent flock with all manner of secular temptations such as English lessons, free travel, and "know how" of various complexions, such as helping those whom they saw as potential converts to start up and run small businesses. One of the consequences has been that many of the traditional Churches have been encouraging the state to curtail the freedom of the foreign religions (and some indigenous new religions), and, at the same time, to protect and enhance their own freedoms by granting them exclusive rights to evangelise, to have free air-time in the media, to provide religious instruction in the schools, and/or to receive financial support (Barker 1997, 2003; Shterin 2000).

Religious Freedom and State Control

Control of religion by the state apparatus can be found to a greater or lesser degree in most, if not all, societies.[21] Sometimes this control has been perfectly benign, even advantageous for the religions concerned; but history provides all too many examples of states attacking the religious freedoms of other countries through invasion and conquest, and of their crushing religions in their own countries (the Soviet regime was but one contemporary example). The repression has been and continues to be both within religions and between religions; the consequences have ranged from irritating byelaws to complete extermination of the targeted groups.

One of the most extreme examples in recent times of a state using religion to control its citizens was the Taliban regime in Afghanistan in the latter half of the 1990s. Women in particular were denied all manner of freedoms, and lived in constant fear of their lives for any minor misdemeanour; they were denied the right to education or to work or even to appear in public unless completely covered by a burqua.²² Under the strict shari'a law (as enforced by some Nigerian states, for example), a person may be executed for converting out of Islam or for blasphemy against it—and those found guilty of adultery may be stoned to death. Elsewhere, the sentences might be less draconian, but still impose severe restrictions on religious freedom. To take one recent innovation, the Gujarat "Freedom of Religious Conversion Bill," passed in March 2003, provides for three years in prison for any conversion ruled to have been by "use of force or by allurement or by fraudulent means." The Bill includes the ruling that any person undergoing conversion must receive prior approval from the head of the district or risk one year's imprisonment.

Historically, Europe has been privy to some of the most brutal scenes of religious persecution throughout the world. Apart from the Crusades, the Inquisition, the Wars of Religion, and the annihilation of various heretical sects, the twentieth century witnessed not only the holocaust, when Jews and Jehovah's Witnesses were murdered in gas chambers along with gypsies, homosexuals and other "undesirables," but also the horrific happenings in the Former Republic of Yugoslavia, when Muslim, Orthodox and Catholic labels were used to intensify the intransigence of the bloody hostilities.²³

In present-day Europe the control of religion is usually more subtle.²⁴ Constitutions nearly always declare that the country supports freedom of religion and most of Europe's states are signatories not only to the United Nations' Universal Declaration of Human Rights and the European Convention on Human Rights and Fundamental Freedoms, but also to numerous other declarations on religious freedom that have been penned by international organisations such as the OSCE.²⁵ None the less, this does not necessarily deter the countries from limiting religious freedoms—particularly in the case of unpopular minority religions. For example, several OSCE participating states have reneged on their commitments by destroying publications belonging to ISKCON (The International Society for Krishna Consciousness).²⁶ Jehovah's Witnesses have been violently attacked in several countries of the Former Soviet Union—the beatings in Georgia have been particularly vehement. They have also been imprisoned for refusing to engage in military service—even in countries, such as Armenia, which has made a commitment to the Council of Europe to end the sentencing of conscientious objectors. It is a curious twist to the cause of freedom that a religion can be viewed as such a threat to society because its members refuse to bear arms.

Although the vast majority of members of the current wave of new religions are no more or less law-abiding than the general population, a small number have been responsible for some appalling tragedies. The first incident to hit the headlines was the mass suicides and murders in Jonestown, Guyana, in 1978 by Jim Jones and his followers in the Peoples Temple; the first to directly affect Europe (especially the French-speaking parts) was the suicides and murders of members of the Solar Temple in 1994/5; and the first to shake the world by directly attacking innocent bystanders completely unrelated to the movement was the release of poisonous gas in the Tokyo underground by Aum Shinrikyo in 1995.

It is not entirely surprising that few would support a religion's freedom to carry out behaviour of this kind. But while some countries consider that their criminal law is sufficient to deal with such actions, other countries have decided that it is necessary to introduce special legislation that distinguishes between religions—and, in the cases of, for example, Austria, France and Belgium, Russia and Armenia, to combat non-traditional religions and sects. France and Belgium set up observatories explicitly for this purpose.[27] Catherine Picard, co-architect of the 2001 French Law for Prevention and Repression of Sectarian Movements, is reported to have said "We need to give judges repressive tools. The law is a response to the evolution of society and the growing importance that sects have in it."[28]

One of the technical problems to which such discrimination gives rise is that of definition. First, it has to be decided what constitutes a "real" religion; secondly, criteria need to be produced for deciding whether a particular religion is an "established," "state," "national," or "traditional" religion—or whether it is a "cult," "sect" or "pseudo-religion" (Barker 1994). Sometimes the distinctions are not made overtly; the criteria for registration can, for example, include length of time in the country or the number of members—which will, of course, tend to discriminate against foreign and relatively new religions, usually putting them at a disadvantage compared to the longer established religions. Another, means of differentiation is simply to name those religions to which special legislation may apply.

Distinguishing between religions for legislative purposes is, however, neither a universal nor a necessary practice. The Netherlands, for example, manages to circumvent the need to evoke an irrelevant and potentially discriminatory definition of religion (let alone a criterion to distinguish between religions), and tax concessions rest on a legal status that is related to the financial (rather than religious) status of the organisation.

Reports

A number of societies have felt the need to commission governmental Reports on new religions. Some of these (Sweden and the Netherlands being examples) concluded that there is no need for distinguishing between new and old religions in law. Others (such as Russia and France) consider the new religions or "sects" to be a potential or actual threat to society. The French and Belgian Reports included lists of, respectively, 172 and 189 sects, the names of nearly all of which were supplied by anti-cult organisations (Gest and Guyard 1995; Duquesne and Willems 1997). The Belgian list included the Quakers, Mormons, Seventh-day Adventists, Opus Dei, a small Jesuit community and the YWCA (though not the YMCA) and various other movements considered perfectly respectable in other societies.[29] The Reports led to protests from a number of quarters, including several scholars of new religions (Introvigne and Melton 1996; Fautre 1998). Belgian and French officials have pointed out that the lists have no legal standing, but the fact that they appeared in governmental Reports would seem to have given permission to those French and Belgian citizens who are so inclined to discriminate against the specified groups, members of which have lost their jobs, been denied the right to buy or rent property or to hire halls for worship, have had their children excluded from certain schools and have, on occasion, been the object of violent attacks, which have included the bombing of Unification and New Acropolis property in Paris (Lheureux 2000; U. S. Dept of State 2000:218, 241–3).

Registration

A requirement for religions to be registered with the state may lead to a reduction of freedoms as it allows the state to keep a close watch on and potentially to control the religion's practices. But registration can also bring special privileges, such as the right to give religious instruction in schools, or to receive various kinds of financial benefits; and not to be registered can incur penalties and restrictions—members may not be able to manifest their religion, and/or to act as a legal entity.[30]

One of the requirements for registration according to the 1997 Russian Federation Federal Law on Freedom of Conscience and on Religious Associations is documental proof that the organisation "has existed over the course of no less than fifteen years on the relevant territory" (Art. 11.5).[31] Those religions that do not succeed in getting registered may be "liquidated,"[32] and at the time of writing several new and not so new foreign religions (such as the Salvation Army and Jehovah's Witnesses) are fighting in the courts in their attempts to avoid liquidation.[33] In October 2002, the President of Belarus signed what has been described as Europe's most repressive Law (Corley 2002). The new law outlaws unregistered religious activity; requires compulsory prior censorship for all religious literature; bans foreign citizens from leading religious organisations; religious education is restricted to faiths that have ten registered communities, including at least one that had registration in 1982; and there is a ban on all but occasional small religious meetings in private homes.

Far more could be written, but space is limited and enough has already been said to indicate that there are many ways in which contemporary states select for special treatment not only religions, but particular religions. This can be to protect their freedoms, but more often it is to curtail the freedoms of both individuals and groups on account of their beliefs as much as, if not more than, because of their actions.

CONCLUDING REMARKS

Several approaches to the subject of religion and freedom have been touched upon. It has been pointed out that individuals may be influenced or persuaded to join, stay in, or leave a particular religion through personal knowledge (or ignorance) or because of the actions of the religion itself, the state, or their immediate social environment. At the group level, religions may reduce or enhance the freedom of their own members and also have considerable influence on the lives of non-members—and they may have their freedoms increased or curtailed by the state or other sections of society. The pressures put upon them can take a variety of forms, ranging from derogatory labels to violent extermination.

It has been argued that the human condition is such that, in the absence of physical restraint or a malfunctioning of the brain, freedom is unlikely to be either totally present or totally absent. Freedom is, rather, a complicated, multifaceted concept, which can be recognised (or hidden) in many guises, and which embraces a number of tensions and apparent paradoxes. Just as what is agoraphobic for one person can be claustrophobic for another, so a social situation experienced as liberating for one individual, can appear stultifying and repressive for another. But social situations, although they are very real and have to be taken into account if we are not to bang our heads against that (socially constructed) brick wall, are also more or less negotiable. There are some situations that we can change—insofar as we understand how we are constrained; there are other situations we cannot change, or which involve a cost greater than we would want to pay.

Maximising freedom requires knowledge of what goes with what, and what the potential consequences might be. Social science cannot tell us what our goals are, but it can try to increase our knowledge and understanding of what information we need to decide whether to accept, change or reject the way things are—and our understanding of how they ought to be.

It has been suggested that the regularities we find in the social sciences are to some degree dependent on our "knowledge" (widely defined) and, thus, to some extent open to change. This means that our work in the sociology of religion can itself contribute to both an increase and a diminution of freedom for those whom we study. If done well,[34] we might empower people to recognise what they can change, and what they cannot change, or what they might make even more intransigent by their attempt to change—which might, of course, be for the promotion of either good or evil.

The other side of the coin is that if we do not point out the complexities, but allow people to interpret our findings in an overly simple direction, without due warning of the processes involved and of the potentially counterproductive consequences of the pursuit of freedom, then we, as sociologists of religion, might be held responsible—to at least some extent—for decreasing their freedom.

Niebuhr appealed to God to help us to have the serenity to accept the things we cannot change, and the courage to change those things we can. This essay has been an appeal to sociologists of religion to contribute to a wisdom that might help us better to recognise and understand the difference.

NOTES

1. Sometimes we say something happened by chance because we don't know why it happened—or it was not planned. This is a different, "softer" kind of chance than what might be an ontological chance—something that happened without any cause and/or for no reason whatsoever.
2. This is not to deny that some kind of chaos does not exist at higher levels of organisation, but it is still debated whether this is an ontological or an epistemological state of affairs—whether the thunderstorm is fundamentally chaotic; or whether we just do not (and probably never can) know enough about the consequences of that butterfly flapping its wings because of the enormous complexity of the connections between the causes. Is, we can ask, the Butterfly Effect "order masquerading as randomness"? (Gleick 1988:22)
3. Some would doubtless prefer concepts such as "correlation" or "constant conjunction" to that of "cause" (Hume 1965:90)—and perhaps they are right, but I do not believe that it is necessary to pursue the status of carnation for our present purposes.
4. In some ways this is similar to the assumption underlying the behaviourism of J. B. Watson (1913) and B. F. Skinner (1972), but at least their social psychology encouraged us to look for the relationships between stimuli and responses, and the intricacies of operant conditioning.
5. It may be noted that this position is in no way similar to fatalism or behaviourism. In some ways it quite the opposite as it assumes that we are all "naturally" free, but the freedom of choice that the convert previously mimed has been lost.
6. See Patrick (1976) for a deprogrammer's own description of the violent methods used in his work to "free" his clients' adult children. It should be added that the more voluntary practices of "exit counseling" and "thought reform counseling" have now almost entirely replaced deprogramming in the West (Giambalvo 1992), though deprogramming involving physical restraint continues in Japan.
7. Now known as the Family Federation for World Peace and Unification.
8. Further analysis indicated, moreover, that it was not those who might, by a number of criteria independent of their joining the movement, be considered particularly weak and/or suggestible who became members (Barker 1984:203).

9. "In Scientology, the individual himself is considered to be the spiritual being—a thetan" (Hubbard 1998:7).
10. The social constructs may be seen as consisting of three inter-related forms: (a) the social structures or institutions that consist of the patterned interactions between individuals in various roles (such as are to be found in the political, economic, educational, welfare, occupational, communication, family—and religious—systems); (b) the cultures that are concerned with our knowledge and understanding of what the world is like; (c) the moral universe, which informs us what the world ought to be like. In all of these areas, groups and individuals in different social positions and with different visions have more or less ability and/or power to negotiate (to innovate, change and/or destroy) various aspects of the ongoing social construction—and those with religious authority have traditionally played a significant role in such processes.
11. Martin Luther "I can do no other." Speech at the Diet of Worms 18 April 1521 (written on his monument at Worms.)
12. "Man is born free and everywhere he is in chains." Opening sentence of Rousseau (1968) chapter one.
13. Church of England Book of Common Prayer, Second Collect, Morning Prayer.
14. Literally "Whose the region, his the religion," meaning that the religion practised by the ruler of a region determines the religion practiced by his or her subjects.
15. It is not being suggested that mare conversion is necessarily beyond the control of the individual. An interesting case in the twentieth century is the mass conversions of Dalits, first out of Hinduism and then into Buddhism, and then, in this century, into Christianity. Interestingly, these have been of a more political than religious nature.
16. This could be seen much earlier, in the classical period and during the Enlightenment, but had been almost entirely confined to the educated or intellectual classes until, perhaps, the French Revolution of 1789.
17. It might be claimed that mine is a somewhat idiosyncratic definition, but it is hard to imagine any definition that would not rely on some order in the physical and social universe.
18. This has also been true for different reasons in Eastern Europe where, as will be discussed in more detail below and in the rest of this special issue, secularism had been imposed by the socialist states with varying degrees of stringency until 1989.
19. Not unrelatedly, Susan Palmer (1994) has argued that one of the attractions for women in new religions is that the movements offer them single rather than multiple roles; they can be accepted as a mother or a lover or a sister (or what have you) instead of constantly having to juggle with the conflicting demands of modern society.

20. In an excellent paper addressing the issue of violence, Edgar Mills (1996:385) argues that normative dissonance increases moral autonomy and serves as a source of restraint upon extreme behaviour in groups.
21. This applies even to countries such as the United States of America with strict separation of Church and State, despite (or, it can be argued, because of) the First Amendment to its Constitution by, for example, forbidding manifestation of religion in public schools. This, of course, takes us into the "right to have freedom from religion" debate.
22. Sally Armstrong, CBC Radio interview October 2001.
23. See the following article by Michael Sells for some gruesome details of the violence in Bosnia-Herzegovina.
24. Indeed, it could be argued that twenty-first century Europe is currently one of the less restrictive regions for religion—which could be to say more about the atrocities that have been occurring elsewhere (particularly in parts of Asia, Africa and the Middle East) than about the liberalism of Europe.
25. The Organization for Security and Co-operation in Europe.
26. Paragraph 16.10 of the Vienna Concluding Document of 1989 states that religious faiths, institutions and organisations should be allowed to produce, import and disseminate religious publications and materials.
27. MILS (la Mission interministerielle de lutte contre les sects/Interministerial Mission in Fight the Sects), succeeded in November 2002 by MIVILUDES (Mission interministerielle de vigilance et de lutte contre les des sectaries/Interministerial Mission to Observe and Fight Deviant Sects) in France; and CIAOSN/IACSSO (Centre d'information et d'avis sure les organisations sectaires nuisibles?/Information and Advice Centre concerning Harmful Sectarian Movements) in Belgium.
28. Joseph Bosco "China's French Connection" *Washington Times* editorial. (10 July 2001).
29. The Swedish Report also included a list of religions in Sweden—but this list was of all the known religions, including both what was then the State Church and five satanic groups (Ingvardsson 1988).
30. See the articles by Schanda, Gunn, and Crnic and Lesjak in this volume.
31. An English translation by Lawrence Uzzel of the Law can be found as Appendix A of Emory International Law Review 12/1 (1998:657–680).
32. Article 14 lays out the grounds for liquidating a religious organisation.
33. On 10 July 2001 the Salvation Amy filed an application with the European Court of Human Rights requesting an intervention to prevent their imminent liquidation by the Russian judicial authorities.
34. By "well" I mean efficiently—providing reliable information that helps us to understand what goes with what, and with what degree in inevitability

this goes with that, under whichever circumstances in whatever situation. In other words, a "good" is good in the sense that he or she is efficient at his or her job and the sense that a "good" corkscrew opens bottles—not that a good corkscrew opens bottles that it is good to have opened.

REFERENCES

Alexander, C. F. 1848. *All things bright and beautiful.* Popular hymn.

Barker, E. 1984. *The making of a Moonie: Brainwashing or choice?* Oxford: Basil Blackwell

———.1994. *But is it a genuine religion?* In *Between sacred and secular: Research and theory on quasi religion,* edited by A. Greil and T. Robbins Greenwich CT; London: JAI Press.

———.1995. The cage of freedom and the freedom of the cage. *Society* 33(3):53–59.

———.1997. But who's going to win? National and minority religions in post-communist society. In *New religions in Central and Eastern Europe,* edited by I. Borowik, 7–44. Krakow: Nomos.

———.2003. Why the cults? New religions and freedom of religion and beliefs. In *Facilitating freedom of religion and belief: Perspectives, impulses and recommendations from the Oslo Coalition,* edited by T. Lindholm, B. Tahzib-Lie, and W. C. Durham. Dordecht NL: Kluwer.

Berger, P. L., and T. Luckmann. 1967. *The social construction of reality: Everything that passes for knowledge in society.* London: Allen Lane.

Bromley, D. G. (ed.) 1988. *Falling from the faith: Causes and consequences of religious apostasy.* Newbury Park, CA: Sage.

Cohn, N. 1970. *The pursuit of the millennium.* Oxford and New York: Oxford University Press.

Corley, F. 2002. Belarus: Europe's most repressive religion law signed today. Oxford: 31 October, Keston Institute, http://www.keston.org.

Davis, D. H. 2002. Explaining the complexities of religion and state in the United States: Separation, integration, and accommodation. In *International perspectives on freedom and equality of religious belief,* D. H. Davis and G. Besier (editors), 169–182. Waco, TX: J. M. Dawson Institute of Church-State Studies, Baylor University.

Duquesne, A., and L. Willems. 1997. Enquete Parlementaire visant d elaborer une politique en vue de lutter contre les practiques illegales des sectes et le danger qu'elles representent pour la societe et pour les personnes, particulierement les mineurs d'age' Brussels: Belgian House of Representatives.

Durkheim, E. 1952. *Suicide: A study in sociology.* London: Routledge & Kegan Paul.

———.1968 (originally published 1915). *The elementary forms of the religious life.* London: George Allen and Unwin.

Fautre, W. (ed.). 1998. *The Belgian state and the sects: A close look at the work of the Parliamentary Commission of Inquiry on sects' recommendations to strengthen the rule of law.* Brussels: Human Rights Without Frontiers.

Gest, A., (President), and J. Guyard (Rappotteur). 1995. *Les sectes en France.* Paris: Assemblee Nationale.

Giambalvo, C. 1992. *Exit counselling: A family intervention: How to respond to cult-affected loved ones.* Bonita Springs, FL: American Family Foundation.

Gleick, J. 1988. *Chaos: The amazing science of the unpredictable.* London: Minerva.

Hubbard, L. R. 1998 (first compiled in 1993). *What is Scientology?* Los Angeles: Bridge.

Hume, D. 1965 (first published 1748) An enquiry concerning human understanding. In *Essential works of David Hume,* edited by R. Cohen, 44–167. New York & London: Bantam.

Ingvardsson, M., et al. 1998. *I God tro: Samhallet och nyandligheten* (In good faith: Society and the new religious movements). Stockholm: Statens offentliga utredninger, Socialdepartementet.

Introvigne, M., and J. G. Melton (eds.). 1996. *Pour en finir avec les sectes: Le debat sur le rapport de la commission parlementaire.* Turin, Paris: CESNUR.

Le Poulet Gauche. 1998. The wars of religion. Part I. History and politics. http://www.lepg.org/wars.htm

Marx, K. 1963 (written 1843–4). Contribution to the critique of Hegel's philosophy of the right. In *Karl Marx: Early writing,* edited by T. B. Bottomore. London: Watts.

Lheureux, N. L. et al. 2000. *Report on discrimination against spiritual and therapeutical minorities in France.* Paris: Coordination des Associations et Particuliers Pour la Liberte de Conscience.

Mills, E. W. 1996. Cult extremism: The reduction of normative dissonance. In *Cults in context: Readings in the study of new religious movements,* edited by L. L. Dawson, 385–396. Toronto: Canadian Scholars' Press.

Palmer, S. J. 1994. *Moon sisters, Krishna mothers, Rajneesh lovers: women's roles in new religions.* Syracuse, NY: Syracuse University Press.

Patrick, T., and T. Dulack. 1976. *Let our children go.* New York: Ballantine.

Popper, K. R. 1961 (first published 1957). *The poverty of historicism.* London: Routledge and Kegan Paul.

Rousseau, J.-J. 1968 (originally written 1740s). *The social contract.* Harmondsworth Penguin

Shterin, M. 2000. New religions in the new Russia. *Nova Religion* 4(2):310–321.

Skinner, B. F. 1972. *Beyond freedom and dignity.* London Pelican.

United States Department of State. 2000. *Annual report: International religious freedom 1999.* Washington, DC: U.S. Government Printing Office.

Wallis, R. 1976. *The road to total freedom: A sociological analysis of Scientology.* London: Heinemann.

Watson, J. B. 1913. Psychology as the behaviorist views it. *Psychological Review* 20:58–77.

Weber, M. 1930. *The Protestant ethic and the spirit of capitalism.* London: Unwin.

Worsley, P. 1968. The trumpet shall sound: A study of "cargo cults" in Melanesia, New York: Schocken.

Wright, S. 1987. *Leaving cults: The dynamics of defection.* Washington, DC: Society for the Scientific Study of Religion.

18

Health, Health Care, and Disability

Have You Checked Your Pension Plan's Funding Level Lately?

Angelo Calvello

The steep decline in the equity markets may have produced some unintended and unfavorable consequences for defined-benefit pension plans. In absolute terms, the total dollar value of most plans probably has been reduced. Perhaps less noticeable, the value of some plans might have declined to the point where it no longer is sufficient to meet the future needs of beneficiaries. In a word, plans might now be underfunded. Healthcare senior financial executives therefore should check the current funding level of their organization's plan.

Until recently, the funding level of a defined-benefit pension plan was rarely a concern. In fact, the buoyant equity markets of the 1990s produced benignly overfunded plans and masked any structural flaws in plans. But the merciless bear market of the past two years has eroded any surplus funding and produced deficits in many plans in the United States. According to the Pension Benefit Guaranty Corporation (PBGC), the government agency that insures the country's 31,000 private pension funds in case they terminate without enough assets to cover their liabilities, "Underfunded pension liabilities for private companies surged to $111 billion at the end of 2001 from $26 billion in 2000."[a] A recent study by Wilshire Associates estimates that more than half of all public pension plans also currently are underfunded.[b] The year-to-date performance of many pension plans makes it likely that this trend will continue through 2002 and into 2003.

Although the trend in employee benefit programs has been toward self-directed defined-contribution plans, the fact remains that the total dollar value of defined-benefit plans in the United States is larger than the value of defined-contribution plans. In June 2002, the top 1,000 defined-benefit plans in the nation had about $3.6 trillion in assets, while the top 1,000 defined-contribution plans totaled about $1.1 trillion.[c] Yet these numbers do not tell the entire story Because the assets of a defined-contribution plan belong beneficially to plan participants, the plan's assets and liabilities must be equal. So $1.1 trillion in assets provides $1.1 trillion in benefits. As the market value of these assets fluctuates, so does the level of benefits.

In contrast, a defined-benefit plan's assets and liabilities are not contemporaneously equal. A defined-benefit plan promises to provide its participants with a predetermined level of retirement benefits in the future. Thus, $3.6 trillion in assets does not equal $3.6 trillion in benefits. Rather, it represents the present value of the plan's future obligations. With the help of an actuary, the plan's sponsor calculates the plan's future pension liabilities and then discounts each future probability-weighted payment back to a present value by using an expected rate of return.

To ensure that assets grow appropriately to meet the plan's future liabilities, the sponsor designs a strategic asset allocation and investment policy that guides the plan's investment of the funds across various asset classes over long periods of time, The objective is to create a portfolio of investments whose aggregate discounted present value consistently equals (or, ideally exceeds) the forecasted liabilities. As long as the aggregate return of the portfolio matches the target return needed to meet the liabilities, then the defined-benefit plan is said to be fully funded. But if the returns to the portfolio come up short, the plan's assets will not be large enough to meet the expected liabilities.

If a plan invests in a diversified portfolio of assets, it inevitably will be underfunded periodically. This underfunding typically is a short-term phenomenon and is not cause for concern. But when a portfolio is exposed to both declining equity markets and low interest rates for a sustained period (the latter causing the present value of its liabilities to soar), the plan could become significantly underfunded. This is precisely the situation financial executives at many healthcare organizations are facing today. Possible negative consequences of this situation include an adverse effect on the organization's balance sheet, disruption of an organization's cash flow, and unwanted scrutiny of the organization by regulatory agencies and investors.

Adverse effect on the balance sheet. If a defined-benefit plan is underfunded, the employer has a legal obligation to make up the shortfall. Just as other companies do, healthcare organizations will fund this contribution out of earnings. Today's environment is causing organizations to use a significant portion of their earnings to make up this shortfall. For example, a recent Wall Street Journal article noted that General Motors had to contribute most of the $3.5 billion generated in its auto business in the second quarter of 2002 to fund its pension plan.[d]

Disruption of cash flow. An underfunded pension plan cannot be ignored; a contribution must be made. But because the need for making a contribution often is not foreseen (especially after the stellar equity market returns of the 1990s), organizations often do not have the required funds built into their budgets. An underfunded plan forces an organization to revise its cash-flow plans and reduce its non-investment-related spending. Moreover, if the company does not make up the shortfall quickly enough, it will have to pay higher PBGC insurance premiums. This increase will come out of the organization's earnings and thus further disrupt planned cash flow.

Unwanted scrutiny. If a plan becomes underfunded, the company is required by law to inform the plan's participants of this change in status—certainly an unpleasant task. This information also inevitably will come to the attention of investors, lenders, vendors, and rating agencies. Also to the point: as part of its credit review process, Standard & Poors has begun asking organizations for the asset valuations and asset allocation details for their defined-benefit plans.

REVIEWING FUNDING STATUS

An underfunded plan is a business problem. Thus, it would be prudent to carefully review the current funding status of the plan and, depending upon the results of this review, take steps to ensure the plan is optimally structured to meet its future liabilities and fiduciary obligations.

First, work with the plan's actuary and legal counsel to review the census information and investment assumptions used to determine the plan's funding status. The latter is especially important. Current losses to the investment portfolio are not the only problem; overly optimistic expectations for future returns that cannot be met negatively impact the funding ratio as well.

Second, evaluate how past and possible future market conditions will affect the organization's balance sheet. The board of directors should be notified of the results of the review, whether good or bad, and informed about remedial action that might improve the likelihood of positively managing the plan's funding level in the future.

Third, evaluate the design, structure, and performance of the plan's investment portfolio. Because the funding status is a factor of the ratio of assets to liabilities, the asset level of the plan directly affects the funding status of the plan. The better the performance of the portfolio, the healthier the plan. Thus, it is important to examine the plan's current strategic asset allocation and investment policy. Is this asset allocation capable of generating a return that can meet the plan's liabilities? Are the asset classes used still the appropriate ones? Would the expected return of the portfolio be improved by, for example, adding new asset classes like alternative investments (such as hedge funds or direct real estate), or reclassifying asset classes (perhaps moving the U.S. equity exposure from rigid value/growth and market capitalization categories to an all-cap mandate)?

Fourth, determine whether the fees charged by all vendors—custodians, managers, and consultants—are truly reasonable and competitive. Burdensome fees often are obscured when performance is strong. But in today's environment, excessive service charges are a needless yet preventable drag on the portfolio. Every dollar paid in fees is a dollar lost in performance.

CONCLUSION

Declining equity markets and low interest rates have negatively impacted the funding level of many defined-benefit plans. By assessing the funding status of their organizations' plans, healthcare financial executives can uncover any funding deficiencies and identify appropriate steps to ensure the plans provide the promised retirement benefits. Following such an assessment, financial executives also need to prepare their organization's board of directors for a possible negative effect on the organization's core business operations and financial resources resulting from any required action.

NOTES

a. Chen K., "Underfunded Pension Liabilities Soared in 2001 to $111 Billion," *The Wall Street Journal,* July 26, 2002, section A, p. 2.

b. Nesbitt, Stephen L, "2002 Wilshire Report on State Retirement Systems: Funding Levels and Asset Allocation," Santa Monica, California: Wilshire Associates Incorporated, August 2002.

c. Feinberg P., "Timber! Falling Funds: Pension Funds Lose at Least $300 Billion in Just Nine Months, *Pensions & Investments,* August 5, 2002.

d. White, Gregory L., "GM Profit Grows, but Pension Costs Worry Investors" *The Wall Street Journal,* online version, July 17, 2002.

19

Population and Urbanization

Unwise Use: Gale Norton's New Environmentalism

David Helvarg

In his State of the Union Address, President Bush called for investing in hydrogen-powered cars. After initial reluctance, the Administration has also implemented Clinton-era proposals to reduce arsenic in drinking water and air pollution from diesel trucks and tractors. And it ordered General Electric to clean up PCB contaminants in the Hudson River. Reporting on what this Administration has done for the environment not only makes for a succinct paragraph but avoids the tedious listing process now required when invoking the ways in which the White House is rolling back a generation of environmental laws, regulations, and treaty commitments.

So when George Bush refers to environmentalists as "Green Green Lima Beans," it's a safe bet he's not engaged in any deep rethinking of policy. Still, as a politician he knows he has to at least appear committed to environmental protection, which is why his political brain, Karl Rove, recently claimed Bush is following in Teddy Roosevelt's environmentalist tradition. That would be the same Roosevelt who condemned "the landgrabbers and great special interests"—the coal, timber, and oil cartels. "The rights of the public to the nation's natural resources outweigh private rights," said T.R.

While the media has portrayed EPA Administrator Christine Todd Whitman as Bush's token environmentalist, Secretary of the Interior Gale Norton has been his real point-woman in promoting "common-sense solutions to environmental policy"—the Republican rhetoric that functions as a pretext for pillage.

When George Bush stood in front of a giant sequoia in California on Earth Day two years ago and spoke of "a new environmentalism for the twenty-first

---

The Progressive, June 2003, v67, i6, p24(6).

© 2003 The Progressive, Inc. Reprinted by permission from *The Progressive,* 409 E. Main St., Madison, WI 53703. www.progressive.org

century" that would "protect the claims of nature while also protecting the legal rights of property owners," Norton was by his side nodding approvingly. A veteran of the small but influential "Wise Use" movement, Norton helped Bush through his environmental tutorial as a Presidential candidate, providing the intellectual arguments that deregulation, devolution, and free markets are the best ways to achieve environmental goals.

Two decades after Ronald Reagan's Secretary of the Interior James Watt used these same arguments to push for the privatization and industrialization of federal lands (among his quotes, "We will mine more, drill more, cut more timber"), Watt's agenda has again become government policy. "Twenty years later it sounds like they've just dusted off the old work," confirms Watt from retirement.

The 1988 Wise Use agenda was written by Watt biographer Ron Arnold, who is executive vice president of the Center for the Defense of Free Enterprise. It called for opening the Arctic National Wildlife Refuge (ANWR) to oil drilling, gutting the Endangered Species Act, opening wilderness lands to energy development, logging, and motorized recreation, and giving management of national parks over to private firms like Disney.

As early as October of 2001, Norton was arguing that opening ANWR to drilling would provide the equivalent of eighty years of Iraqi oil imports (pre-invasion) to the United States. She's also pursuing energy development, logging for "forest health" and motorized recreation on public lands, mountain-top removal for coal mining in Appalachia, and captive breeding of endangered species in lieu of habitat protection. She's reversed a plan that would have banned snowmobiles from Yellowstone and Grand Teton National Parks, she is limiting the amount of land set aside for wilderness protection, and she is moving forward with a plan to begin "outsourcing" National Park Service jobs to private firms.

"I wish we could take credit for that but we can't," admits Ron Arnold. "Sometimes you just put something out there long enough and it gets picked up, despite what you do."

There was a brief moment, in the earliest days of the Bush Administration, when it appeared the White House might balance its thirst for oil with a nod toward wilderness protection by naming John Turner as Secretary of the Interior. Former head of the U.S. Fish and Wildlife Service under Bush Senior, Turner is also a Wyoming fly-fishing buddy of Dick Cheney's. He comes from the clipped moderate wing of the Republican Party that sees conserving nature as part of the conservative tradition.

But before Turner's name could be put forward, the White House was flooded with angry faxes and e-mails organized by the American Land Rights Association's Chuck Cushman, a Wise Use firebrand who maintains an e-mail action list of thousands of property rights, mining industry, and National Cattlemen's Beef Association members. He warned his listserv, "Turner's long-standing relationships with the Rockefeller Family Foundation and their financing of the environmental left" would lead to a green takeover of the Department of the Interior by "Land-Grabber Turner."

Backstopped by his Washington lobbyist Mike Hardiman at the American Conservative Union, Cushman and the Wise Use fringe were able to do what they do best: sabotage a potentially eco-friendly initiative, in this case by scaring the Bush White House into thinking it might lose some of its core constituents on the hard right.

"They caved. They blinked. Cheney's probably angry at us but who cares," says Arnold. "Norton is a friend."

And an old one at that. Norton got her start at Watts Mountain States Legal Foundation in Denver, which billed itself as the "litigation arm of Wise Use." While there, she argued that the government should pay financial compensation whenever environmental laws limited a developer's real or potential profits.

This argument is based on the Fifth Amendment, which states, "No person shall be . . . deprived of life, liberty, or property, without due process of the law: nor shall private property be taken for public use without just compensation." Today, when the government condemns land to build a highway (or a Texas baseball stadium) the Fifth Amendment guarantees that the landowner be paid market value for his or her lost property.

Yet, in 1887, when a Kansas beer brewer argued a prohibition law in his state was also a takings under the Fifth Amendment, the Supreme Court ruled against him, stating that, "a government can prevent a property owner from using his property to injure others without having to compensate the owner for the value of the forbidden use."

But Norton has argued for a major shift in the takings jurisprudence. "We might even go so far as to recognize a homesteading right to pollute or to make noise in an area. This approach would eliminate some of the theoretical problems with defining a nuisance," she wrote in the *Harvard Journal of Law and Public Policy* back in 1990. She later recanted her "right to pollute" phrase during her Senate confirmation hearing.

In 1998 Norton became co-chair of the Coalition of Republican Environmental Advocates. Dedicated to "free market environmentalism," the coalition included auto, coal mining, and developer lobbyists. Traditional Republican environmentalists like the late Senator John Chafee of Rhode Island refused to join.

In 1999 Norton, now working as a lawyer representing the lead industry, became part of the team advising candidate Bush on developing a conservative "environmentalism for the twenty-first century." Among those working with her was David Koch of Koch Industries, which in 2000 paid a $35 million fine for oil pollution in six states, as well as Lynn Scarlett from the libertarian Reason Foundation of Los Angeles. Scarlett was also a senior fellow at Montana's Foundation for Research on Economics and the Environment (FREE), which lived up to its acronym by holding a series of all-expenses-paid property-rights "seminars" for federal judges at a Montana dude ranch.

Norton also spent time as a fellow at the Political Economy Research Center (PERC), Montana's other property-rights think tank. David Koch is a major funder of both FREE and PERC. With his family's $23 billion fossil fuel fortune, Koch also bankrolls a hornet's nest of D.C.-based free market think tanks, including the Heritage Foundation, the Cato Institute, the Competitive Enterprise Institute, and Citizens for a Sound Economy, which all advocate for the "new environmentalism" of deregulation.

"The last three decades is what I call the old environmentalism," Scarlett says. Scarlett is now Norton's assistant secretary for policy, management, and budget, and since Norton likes to keep a low profile, Scarlett has become Norton's emissary to the public.

"The old environmentalism tended to rely on the four 'P's: prescription, telling you how you're supposed to do things; process, a focus on the permit you need to 'pass go' rather than the result; punishment, as a way to motivate action, and a piecemeal approach to air, land, and water," Scarlett explains. "We're not getting rid of regulation, but shifting the emphasis, extending a hand to work with, not against, landowners. We think real sustainability has to be about what we call Cooperative Conservation, about engaging people."

The Department of the Interior is now packed with people previously engaged in the employ of industry. The departments Deputy Secretary Steve Griles is a former mining and oil lobbyist. The Senate, before confirming him, made him sign an agreement that he wouldn't meet with his former clients. Nevertheless, Griles has gone on to meet with his former clients to discuss new rules that allow the dumping of coal mining waste in Appalachian rivers and a massive coal-bed methane drilling project in Wyoming, according to The Washington Post. On April 7, Senator Joe Lieberman, Democrat of Connecticut, asked the Interior Department's inspector general to investigate Griles following a report by the Associated Press that Griles also attended meetings on offshore leases in which his former clients had significant interests.

"These reports raise numerous, troubling questions about whether the deputy secretary has successfully avoided conflicts of interest, or the appearance of conflicts," said Lieberman.

Eric Ruff, director of communications at the Department of the Interior, disputes the facts of *The Washington Post* account and says Griles "has followed the highest ethical standards of the department."

Norton's special assistant on Alaska is a former oil lobbyist, her assistant secretary for water and science is a former mining lawyer who has called for the abolition of the Endangered Species Act, and her solicitor is from the Cattlemen's Beef Association, where he lobbied for cheap grazing fees on Interior lands.

With those kinds of built-in conflicts of interest, even Norton's collaborative rhetoric and her attempts at outreach (including a letter to actor/environmentalist Robert Redford in which she pitched their shared love of zoo-bred condors) haven't been enough to prevent a series of in-house scandals.

Ironically her biggest controversy is an inherited one, the century-old accounting mess at the Bureau of Indian Affairs. American Indians claim the federal government has squandered billions in oil, gas, and timber royalties from their lands and are suing to reclaim the money.

A more recent scandal that has also angered California Indian tribes involves the death of more than 35,000 salmon on the lower Klamath River, attributed to low water flows after the U.S. Bureau of Reclamation diverted water north to Oregon farmers.

In the spring of 2001, alfalfa, hay, and potato farmers marched through the streets of Klamath Falls, Oregon, and illegally opened dam gates to protest a federal decision cutting their irrigation water to guarantee protection for endangered suckerfish and coho salmon runs. Like the snail darter and spotted owl before it, the suckerfish quickly became the poster-animal for anti-Endangered Species Act pundits from talk radio's Rush Limbaugh to the editorial writers at *The Wall Street Journal*. What few of these conservative critics were willing to acknowledge was that the water crisis was precipitated by the worst drought to hit the Northwest in over a century, a drought that, like the region's forty-six shrinking glaciers, is likely linked to fossil-fuel-enhanced climate change. "1994 was the last substantial rain we had," Ryan Kliewer, a young fourth-generation farmer who marched in the protests, told me.

Under pressure from the White House, Norton's Bureau of Reclamation slashed the river flow, returning much of the water to the irrigators, despite a report from a team of federal scientists warning this would place the coho salmon in serious jeopardy (along with the commercial fishermen and Indian tribes who depend on them). Last October, Representative Mike Thompson and a group of California protesters dumped 500 pounds of dead, rotting Klamath salmon on the front steps of the Department of the Interior, accusing Norton of a massive cover-up.

This wasn't the first time Norton's been accused of ignoring or suppressing government scientists.

In the fall of 2001, she had to explain why, in a letter to the Senate Committee on Energy and Natural Resources, she'd altered scientific data to make it appear that oil operations in ANWR would not harm hundreds of thousands of migratory caribou, when her own Fish and Wildlife Service had provided her with data suggesting they would.

"We did make a mistake. We will take steps to clarify and correct that," she told reporters in explaining one of the numerous discrepancies.

Norton also concluded that drilling the Arctic wouldn't violate an international treaty that protects polar bears. The Fish and Wildlife Service, which has twice issued reports stating that drilling poses a threat to the bears, was directed "to correct these inconsistencies" (in line with Norton's position).

Bush also signed off on an Army Corps of Engineers proposal that makes it easier for developers and mining companies to dredge and fill America's wetlands through a "general permitting" process that is rarely if ever challenged. Again, Norton failed to forward comments from her Fish and Wildlife Service to the Army Corps, even though the service had written that the proposed policy change would result in "tremendous destruction of aquatic and terrestrial habitat."

Norton's top aides are now actively monitoring career staffers to make sure that scientific assessments don't conflict with their pro-business political goals, according to whistleblowers within the department who've been in touch with reporters and environmental groups. As a result, morale among Interior field scientists is said to be falling faster than a wing-shot condor.

The latest suppressed study (which surfaced in *The New York Times* on January 31) came from the National Park Service. Today, the largest constituency organized to open up parks and wilderness areas to roads and development is no longer Wise Use loggers and resource industry employees but suburban owners of motorized dirt bikes, ATVs, snowmobiles, and personal watercraft. While millions of Americans are having a love affair with fast, loud off-road vehicles, their owners are creating major user conflicts with tens of millions of other outdoor recreationists who enter wilderness areas believing they've left the noise and pollution of the freeway behind.

Air pollution from winter snowmobilers and harassment of buffalo and other wildlife got so bad in Yellowstone National Park that the Park Service decided to phase out the activity. But Norton's Department of Interior—in response to a suit from the International Snowmobile Manufacturers Association—announced it would reassess the rule-making process, this despite 360,000 e-mails and letters, 80 percent of which supported banning the machines.

"I think the national environment groups have turned on recreation with a vengeance now that they've driven the commodity users off the [public] lands," argues Bill Horn, Washington counsel for the Snowmobile Manufacturers. "Our conversations with the Secretary [of Interior Norton] show her greater appreciation of public recreation on public lands. We need these places for everyone's enjoyment—not just to have these scientists go in there and create biospheres under glass."

In November, the Bush Administration proposed a cap of 1,100 snowmobiles a day, up from the present average of 815, arguing that a new generation of machines will be quieter and less polluting. At the same time, according to the *Times*, the Park Service concluded an internal report (not made public) that found banning the machines was the best way to protect the park's air quality and wildlife and the health of visitors and employees. Still, thousands of snowmobiles continued to make their runs through Yellowstone and Grand Teton this winter and will continue to do so, at least for the remainder of the Bush Administration.

The next big push at the Interior Department is likely to be the privatization of thousands of National Park Service jobs, including the entire corps of park scientists. The government-wide "competitive sourcing" initiative, being promoted by Bush and his Office of Management and Budget, targets as many as 850,000 jobs in the largely unionized federal workforce to be replaced by private contractors, including up to 11,524 out of 16,470 National Park Service jobs.

Fran Mainella, director of the Park Service, sent an e-mail to her employees at the end of January assuring them that while 70 percent of their jobs were being studied to see if they were "inherently governmental" functions, that "70 percent has never been used as a measuring stick for privatizing National Park Service jobs." Under present plans, 15 percent of Park Service jobs will be outsourced by 2004, which will, according to Scarlett, "help tap professional tools with better delivery of services to the public, and new skills and technologies and discipline."

"The plan is designed to meet ideologically set goals and will rip apart the fabric of the agency," counters Jeff Ruch, executive director of Public Employees for Environmental Responsibility, a Washington-based activist organization that's represented a number of Park Service workers. Ruch claims that replacing park scientists will lead to "private consulting firms telling the Park Service what it wants to hear in order to get their contracts renewed."

With the Department of the Interior also promoting recreational user fees, corporate sponsorship of park activities, and partnerships for bio-prospecting (companies looking to develop new drugs from microbes, plants, and animals) in the parks, one can start to imagine Smokey the Bear recast as ComCast Bear, Arches National Park as Golden Arches National Park, or hip ads promoting Yellowstone washed jeans. Certainly the Wise Use vision of park management given over to private firms, "with expertise in people-moving such as Walt Disney," seems consistent with Norton's policies.

As does the end of wilderness itself. On the evening of Friday, April 11, after Congress had recessed for spring break, the Department of the Interior quietly released a statement announcing that, after almost forty years of scenic and biologically important habitat protections under the Wilderness Act, it would no longer seek any additional wilderness designations on public lands. This potentially opens up some 250 million acres of federal rangelands and Western mountains to mining, drilling, road construction, and other forms of development.

Still, by keeping a low profile and being a team player, by talking about her love of hiking, condors, and conservation while promoting the White House energy plan to open millions of acres of public lands to mining and drilling, by choosing her audiences and never second-guessing her boss, Norton has remained a relatively noncontroversial figure in the Administration. Despite her early portrayals by environmentalists as "James Watt in a skirt," she has shown far more political acumen than the man who once bragged of a commission on which he had a black, a woman, "two Jews, and a cripple."

Rather then openly attack environmental laws like the Endangered Species Act (which she has argued is unconstitutional), Norton has used the regulatory process to ease up on industry and the administrative process to crack down on agency professionals who disagree with her.

Of course, with the Republicans now in charge of both the White House and Capitol Hill, the impulse to overreach and go for a more fundamental realignment of broadly popular environmental laws may prove highly tempting.

For environmentalists to move beyond a purely defensive posture, however, it will take more than waiting on a Republican gaffe or alarmed direct mail solicitations showing nursing caribou on the tundra.

The environmental movement has to address the links among oil, energy, climate change, landscape, and security. This will mean dealing with the Middle East wars, as well as Norton's drilling permits in the Gulf of Mexico and Alaska. It will mean committing to a level of national and global politics that could prove highly partisan, complex, and intense.

20

Collective Behavior, Social Movements, and Social Change

A Retrospective on the Civil Rights Movement: Political and Intellectual Landmarks

Aldon D. Morris

This review provides an analysis of the political and intellectual contributions made by the modern civil rights movement. It argues that the civil rights movement was able to overthrow the Southern Jim Crow regime because of its successful use of mass nonviolent direct action. Because of its effectiveness and visibility, it served as a model that has been utilized by other movements both domestically and internationally. Prior to the civil rights movement social movement scholars formulated collective behavior and related theories to explain social movement phenomena. These theories argued that movements were spontaneous, non-rational, and unstructured. Resource mobilization and political process theories reconceptualized movements stressing their organized, rational, institutional and political features. The civil rights movement played a key role in generating this paradigmatic shift because of its rich empirical base that led scholars to rethink social movement phenomena.

INTRODUCTION

What would America be like if we could turn back the clock to 1950? One thing is certain, the pre-civil rights movement era would stand in stark contrast to the America that currently exists just two years before the new millennium. In terms of race relations, the contrast is so sharp that we are justified to speak of a pre- and post-civil rights movement period. The civil rights movement is clearly one of the pivotal developments of the twentieth century. The task of this chapter is to elucidate the factors that made this movement such an important force in America and abroad. The chapter also concerns itself with how the civil rights movement has affected the theoretical developments in the field of social movements. The first concern is to discuss why the civil rights movement was necessary in the first place.

The Jim Crow regime was a major characteristic of American society in 1950 and had been so for over seven decades. Following slavery it became the new form of white domination, which insured that Blacks would remain oppressed well into the twentieth century. Racial segregation was the linchpin of Jim Crow, for it was an arrangement that set Blacks off from the rest of humanity and labeled them as an inferior race (see Morris 1984:2). Elsewhere I characterized Jim Crow as a tripartite system of domination (Morris 1984) because it was designed to control Blacks politically and socially, and to exploit them economically. In the South, Blacks were controlled politically because their disenfranchisement barred them from participating in the political process. As a result, their constitutional rights were violated because they could not serve as judges nor participate as jurors.

Economically Blacks were kept at the bottom of the economic order because they lacked even minimal control over the economy. Throughout the first half of the twentieth century, most rural Blacks worked as sharecroppers and hired hands where they were the victims of exploitation because of the unequal economic arrangements they were forced to enter. As Blacks migrated to Southern and Northern cities they found that their economic status did not change radically; in these settings they were forced onto the lowest rungs of the unskilled wage sectors. Therefore, "in 1950 social inequality in the work place meant that nonwhite families earned nationally 54% of the median income of white families" (Morris 1984:1).

The social oppression Blacks experienced prior to the civil rights movement was devastating. The Jim Crow system went to great lengths to impress on Blacks that they were a subordinate population by forcing them to live in a separate inferior society. Moreover, the fact that Blacks had to use separate toilets, attend separate schools, sit at the back of buses and trains, address whites with respect while being addressed disrespectfully, be sworn in on different bibles in the court room, purchase clothes without first trying them on, pass by "white only" lunch counter seats after purchasing food, and travel without sleep because hotels would not accommodate them—all these—resulted in serious psychological damage.

The role that violence and terror played against Blacks during the Jim Crow period has been documented. John Hope Franklin (1967) detailed the violence that white supremacist groups, including the Ku Klux Klan and the Knights of the White Camellia, heaped upon African Americans. He wrote that such groups "used intimidation, force, ostracism in business and society, bribery at the polls, arson, and even murder to accomplish their deeds" (Franklin 1967:327). The lynch rope was one of the most vicious and effective means of terror used against Blacks. In a painstaking analysis of lynching in one Southern state, Charles Payne (1994) wrote that "between the end of Reconstruction and the modem civil rights era, Mississippi lynched 539 Blacks, more than any other state. Between 1930 and 1950—during the two decades immediately preceding the modem phase of the civil rights movement—the state had at least 33 lynchings" (1947:7).

As late as the 1950s the Jim Crow regime remained firmly intact, effectively oppressing the Southern Black population. Earlier in the century W.E.B. DuBois had predicted that the problem of the twentieth century was the problem of the color line (DuBois 1903). By the 1950s racial realities suggested that the color line was not likely to undergo substantial change during the twentieth century. As late as the 1940s, the majority of white Americans still held racist views that squarely supported white supremacy. Larry Bobo summarized the evidence:

The available survey data suggest that anti-Black attitudes associated with Jim Crow racism were once widely accepted. The Jim Crow social order called for a society based on deliberate segregation by race. It gave positive sanction to anti-Black discrimination in economics, education, and politics. . . . All of this was expressly premised on the notion that Blacks were the innate intellectual, cultural and temperamental inferiors to whites (Bobo 1997:35).

In their heyday, systems of domination often appear unshakable. By 1950 the Jim Crow regime appeared to rest on a solid foundation of white supremacy capable of enduring indefinitely.

Beneath the placid appearance of permanent dominance there may exist substantial resistance. For the Black population such resistance had gathered steam long before 1950. It is often the case that students of protest overlook resistance until it becomes highly visible and threatens dominant interests. The inability to grasp such "subterranean" forces is more pronounced when the oppressed population is perceived as inferior and incapable of generating the agency required to transform domination. As I discuss later, this certainly was the case for African Americans. We now have nuance studies documenting the continuity of Black protest and Black insurgent ideologies. Vincent Harding's (1983) study, *There Is A River*, details the numerous protests that African Americans initiated throughout the slave period and the radical visions associated with these struggles. George Fredrickson's (1995) *Black Liberation*, does a masterful job documenting how African Americans developed radical ideologies from the nineteenth century to the present that have guided their struggles and played significant roles in liberation ideologies on the continent of Africa. These studies make clear that African Americans have long possessed a protest tradition that is continually refashioned and interjected into new rounds of struggles.

Early years in the twentieth century African Americans launched protests directly attacking racial inequality. Between 1900 and 1906 Southern Blacks developed boycott movements against Jim Crow streetcars in most major cities of the South (Meier & Rudwick 1976:267–89). By the turn of the twentieth century Black women had organized local and national clubs through which they relentlessly fought both for the overthrow of Jim Crow and for women's rights (Collier-Thomas 1984:35–53). The National Association for the Advancement of Colored People (NAACP) was founded in 1909–1910. This was an important development because the NAACP was the first national protest organization organized specifically to attack the Jim Crow regime and racial inequality. It immediately began attacking the legal basis of racial subordination during the Jim Crow era (McNeil 1983, Tushnet 1987). The NAACP would win major legal cases against racial segregation throughout the first half of the twentieth century especially with regards to segregated schools.

Major developments occurred in the 1920s that challenged entrenched ideas of white supremacy and Black inferiority. The Garvey movement, organized in 1920, rapidly became the largest mass movement Black America had ever produced. Its main message was that Black people, Black culture, Black history and Africa were noble and that Black people had created great civilizations that rivaled Western civilization on every front. Garvey preached that Blacks should return to Africa. This praising of things Black flew directly in the face of white hegemonic beliefs. Large numbers of African Americans were receptive to this message. Otherwise the Garvey movement could never have developed into a major mass movement.

The Harlem Renaissance of the 1920s was a major literary movement that carried a similar message. This movement produced what has come to be characterized as protest literature. It aimed at creating a "New Negro" who was proud of her Black heritage and prepared to fight for Black liberation. This protest theme was clearly represented in lines of Claude McKay's poem, "If We Must Die" where he declared, "Oh, Kinsmen! We must meet the common foe! Though far outnumbered, let us show us brave/ And for their thousand blows deal one deathblow!" (McKay 1963:31).

During this period Black protests accompanied the proliferating Black protest literature. Jaynes & Williams found that "between 1929 and 1941, Northern Blacks organized a series of 'don't buy where you can't work' campaigns in which white owned ghetto businesses were boycotted unless they agreed to hire Blacks" (Jaynes & Williams 1989). But it was a movement organized just a decade prior to the explosion of the modern civil rights movement that was the clearest harbinger of things to come. In the early 1940s A. Philip Randolph organized the March on Washington Movement (MOWM). Randolph had become convinced that a mass nonviolent movement of African Americans was the central force needed to overthrow racial inequality. Throughout 1941, Randolph, Black leaders, and organizers across America worked to build a Black nonviolent direct action movement (Garfinkel 1969). It was named the March on Washington Movement because Randolph decided to target racial discrimination in the defense industries by marching on Washington and the White House by the thousands. Such a mass march, he reasoned, would embarrass the nation and President Roosevelt who were in the midst of fighting racism abroad during World War II. On the eve of the March Randolph and Black leaders throughout the nation had organized thousands of Blacks who were prepared to march on Washington. The March never occurred because Roosevelt suspected that it was a real possibility that thousands of Blacks would march on his White House. On June 25, 1941 he issued an Executive Order that banned racial discrimination in the nation's defense industries. Thus, the very threat of protest by a massive Black nonviolent direct action movement bore fruit in the early 1940s.

Two sharply contrasting developments occurred in the early 1950s that severely threatened the Jim Crow system. They were the 1954 Supreme Court ruling in the Brown vs. Board of Education case and the lynching of Emmett Till in 1955. In the Brown case the NAACP won a major Supreme Court ruling that declared racially segregated schools unconstitutional (Kluger 1975). This ruling literally swept away the legal grounds on which Jim Crow stood. African Americans were filled with hope by the ruling, believing that finally it was realistic for them to believe that legal racial segregation was on its deathbed. Morehouse College President, Benjamin Mays, captured the Black reaction when he stated "we all underestimated the impact of the 1954 decision . . . people literally got out and danced in the streets. . . . The Negro was jubilant" (Mays Interview 1978). The Southern white power structure had a completely different reaction. They rebuked the Supreme Court for its decision, vowing never to integrate the schools nor to dismantle the Jim Crow regime. The Brown ruling caused sharp battle lines to be drawn between the white South and African Americans.

In August of 1955, Emmett Till, a fourteen year old Black male from Chicago, was lynched in Money, Mississippi for whistling at a white woman (Whitefield 1988). The crime was extremely brutal, and it was a reminder to the Black community that whites would utilize all means, including murder, to uphold Jim Crow. An all-white jury exonerated Till's murderers of the crime. The black response to Till's lynching differed from the usual pattern. Till's mother and the Black press generated national publicity pertaining to the gross injustice of the lynching. Because of the widespread attention this lynching received, the brutality and raw racism of the Jim Crow regime were placed on a national stage where it was debated and denounced.

The generation of young Blacks who would lead the student wing of the modern civil rights movement was coming of age precisely at the time of Till's lynching. This murder played an important role in radicalizing them. They were shocked at the brutality of the crime and outraged when the murderers were allowed to go free by the white judicial system. Many of them began embracing ideas of activism because they themselves felt vulnerable. They were well aware that the white community and many adults within the black community refused to fight for justice. Thus, the Till lynching pushed them toward political activism (see Moody 1968, Ladner 1979, Whitfield 1988). The hope generated by the Brown Ruling and the outrage caused by Till's lynching, helped set the stage for the emergence of the modem civil rights movement.

Structural Prerequisites of the Civil Rights Movement

Oppressed groups are not always in a position to generate change through social protest. Favorable social conditions play an important role in creating the circumstances conducive to protest. Social movement scholars (McAdam 1982, Tarrow 1994) have asserted that social protest is more likely to occur if there exists a favorable political opportunity structure. Such a structure had developed prior to the rise of the modem civil rights movement. As McAdam has argued (1982), by the time the civil rights movement began to unfold, Blacks had amassed a new level of political power because of the Northern Black vote. That vote could be used to push a Black agenda, especially given its emerging importance in presidential elections.

The politics of the Cold War was an additional factor making Black protest a viable option. The United States and the Soviet Union were locked in an intense battle to win over newly independent Third World countries, especially those in Africa. The issue of American racism was an impediment to an American foreign policy bent on persuading African nations to align themselves with America. Racism and democracy were opposing ideologies, and Black leaders were aware that America's treatment of Blacks could be a stumbling block in America's quest to become the major superpower. Wide-scale Black protest, therefore, stood a good chance of exposing the contradiction between racism and democracy.

The coming of age of modem communication technologies in the 1950s and early 1960s was another development that could be exploited by a Black protest movement. The widespread use of television was a case in point. As early as 1958, over 83% of American households owned television sets (Sterling & Kittross 1978). These technologies were capable of providing a window through which millions could watch Black protest and become familiar with the issues it raised. Likewise, by the early 1960s communications satellites were launched into orbit. This development made it possible for Black protest to be viewed globally, thus enhancing its ability to affect the international arena.

A development internal to the Black community also increased the probability that widespread protest could be launched and sustained. This was the great migration that occurred during the period of the two World Wars and that continued throughout the 1950s. During this period large numbers of Blacks moved to cities in both the South and the North. This migration led to institution building especially within the Black Church and community organizations. These were the kinds of institutions through which protest could be organized and supported. The urban setting also provided the Black community with dense social networks through which social protest could be organized rapidly. In short, by the 1950s the Northern Black vote, the politics of the Cold War, the rise of modern communication technologies, and Black mass migration constituted favorable social conditions conducive to the rise of a massive Black movement.

The existence of favorable conditions does not guarantee that collective action will materialize. Agency is required for such action to occur. People must develop an oppositional consciousness that provides them with a critique of the status quo and reasons to believe that acting collectively will lead to change. They must develop the willingness to make sacrifices that may endanger their physical well being or cause them to lose jobs. They must be willing to devote time to collective action that may cause them to neglect or curtail other important but routine activities. Creativity is crucial to social movements because they require new ways of doing things and they thrive on innovations. People who participate in movements have to place themselves in learning situations where they can be taught to act creatively. Structural prerequisites may be conducive to collective action, but without human agency such conditions will not even be recognized, let alone exploited.

THE MODERN CIVIL RIGHTS MOVEMENT

The impact of the civil rights movement on race relations and the nation's social fabric has been monumental. This pivotal movement has had significant influence on social movements in a wide array of countries. The intent here is not to provide a detailed account of the modern civil rights movement. Such accounts are available in the vast literature that has emerged over the last twenty years (e.g. Carson 1981, Sitkoff 1981, McAdam 1982, Blumberg 1984, Morris 1984, Garrow 1986, Bloom 1987, Fairclough 1987, Branch 1988, Lawson 1991, Robnett 1997). The purpose here is to present an analysis of why this movement by a relatively powerless group was able to overthrow the formal Jim Crow system and how it became a model for other protest movements here and abroad. My analysis draws heavily from the previously mentioned literature.

The most distinctive aspect of the modern civil rights movement was its demonstration that an oppressed, relatively powerless group, can generate social change through the widespread use of social protest. For nearly two decades, this movement perfected the art of social protest. The far ranging and complex social protest it generated did not emerge immediately. Rather it evolved through time making use of trial and error.

By the mid 1950s Southern Black leaders had not yet fully grasped the idea that the fate of Jim Crow rested in the hands of the Black masses. Even though protest against racial inequality occurred throughout the first half of the century, it tended to be localized and limited in scope. With the exception of the Garvey Movement and A Philip Randolph's March On Washington Movement (MOWM), the mass base of the prior protests was too restricted to threaten the Jim Crow order. Both the Garvey and MOWM movements had limited goals and were relatively short lived. By 1950 the legal method was the dominant weapon of Black protest, and it required skilled lawyers rather than mass action. The legal method depended on the actions of elites external to the Black community whereby Blacks had to hope that white judges and Supreme Court justices would issue favorable rulings in response to well-reasoned and well-argued court cases.

The 1955 Montgomery, Alabama, year-long mass-based bus boycott and the unfolding decade of Black protest changed all this. These developments thrust the power capable of overthrowing Jim Crow into the hands of the Black community. Outside elites, including the courts, the Federal Government, and sympathetic whites, would still have roles to play. However, massive Black protest dictated that those roles would be in response to Black collective action rather than as catalysts for change in the racial order. A decisive shift in the power equation between whites and Blacks grew out of the struggle to desegregate Alabama buses.

Out of Montgomery came the Montgomery Improvement Association (MIA), the first highly visible social movement organization (SMO) of direct action protests. It was both church based and structurally linked to the major community organization of the Black community. A young Black minister, Martin Luther King, Jr., who came to personify the role of charismatic leader, was chosen to lead the movement. Finally, it was the boycott method itself that shifted power to the Black masses, for it required that large numbers of individuals engage in collective action. The requirement that the boycott adopt nonviolent direct action was crucial, for it robbed the white power structure of its ability to openly crush the movement violently without serious repercussions. The mass media, especially television, radio, and the Black press, as well as communication channels internal to the Black community, proved capable of disseminating this new development across the nation. A ruling by the Supreme Court, that declared bus segregation in Alabama unconstitutional, sealed the victory for the movement. It was clear, however, that the ruling was in response to Black protest and that Jim Crow could be defeated if the lessons of the Montgomery movement could be applied in movements across the South which targeted all aspects of racial segregation.

The Montgomery bus boycott revealed that large numbers of Blacks—indeed an entire community—could be mobilized to protest racial segregation. The Montgomery boycott demonstrated that protest could be sustained indefinitely: the boycott endured for over a year. The boycott revealed the central role that would be played by social organization and a Black culture rooted in a protest tradition, if protests were to be successful. The Black Church, which had a mass base and served as the main repository of Black culture, proved to be capable of generating, sustaining, and culturally energizing large volumes of protest. Its music and form of worship connected the masses to its protest tradition stemming back to the days of slavery.

For more than a decade, such movements did occur across the South. Numerous local communities throughout the South, supported by the national

Black community and sympathetic whites, perfected the use of mass-based nonviolent direct action. The disruptive tactics included boycotts, sit-ins, freedom rides, mass marches, mass jailings, and legal challenges. The 1960 student sit-ins at segregated lunch counters were especially important because within a month these protests had become a mass movement, which spread throughout the South and mobilized an important mass base. That base initially consisted largely of Black college students but also came to include high school and elementary students. The sit-in tactic was innovative because other tactics spun off of it, including "wade ins" at segregated pools, "kneel-ins" and "pray-ins" at segregated churches, and "phone-ins" at segregated businesses.

The sit-in movement was also critical because it led to the formation of the Student Nonviolent Coordination Committee (SNCC), which became a social movement organization of students through which local communities were organized and mobilized into social protest (Morris 1984). This organization increased the formal organizational base of the movement, which already included the Southern Christian Leadership Conference (SCLC), the Congress of Racial Equality (CORE), the NAACP, and numerous local movement organizations. The sit-in movement and SNCC constituted the framework through which large numbers of white college students came to participate in the civil rights movement.

It was during the early to mid 1960s that the modern civil rights movement became the organized force that would topple Jim Crow. In this period highly public demonstrations occurred throughout the South and came to be increasingly strengthened by Northern demonstrations that were organized in support. These protests created a crisis because they disrupted social order and created an atmosphere that was not conducive to business and commerce in the South (Bloom 1987). They often caused white officials to use violence in their efforts to defeat the movement. By this time Martin Luther King Jr. had become a national and international charismatic figure. As James Lawson put it, "Any time King went to a movement, immediately the focus of the nation was on that community. . . . He had the eyes of the world on where he went. And in the Black community, it never had that kind of person. . . . It gave the Black community an advantage [that] it has never had" (Lawson 1978).

The intensity and visibility of demonstrations caused the Kennedy Administration and the Congress to seek measures that would end demonstrations and restore social order (Morris 1984, Schlesinger 1965). The demonstrations and the repressive measures used against them generated a foreign policy nightmare because they were covered by foreign media in Europe, the Soviet Union and Africa (see Fairclough 1987, Garrow 1986). As a result of national turmoil and international attention, the Jim Crow order was rendered vulnerable.

The Birmingham, Alabama, confrontation in 1963 and the Selma, Alabama, confrontation in 1965 generated the leverage that led to the overthrow of the formal Jim Crow order. In both instances, the movement was able to generate huge protest demonstrations utilizing an array of disruptive tactics that caused social order to collapse. The authorities in each locale responded with brutal violence that was captured by national and international media. The Birmingham movement sought to dismantle segregation in Birmingham and to generate national legislation that would invalidate racial segregation writ large. Because the Birmingham confrontation was so visible and effective it engendered additional protests throughout the South. Within ten weeks following the Birmingham confrontation, "758 demonstrations occurred in 186 cites across the South and at least 14,733 persons were arrested" (Sale 1973:83).

The goal of the 1965 Selma movement was to generate the necessary pressure that would lead to national legislation that would enfranchise Southern Blacks. The Selma confrontation commanded the same level of attention as the confrontation in Birmingham. Over 2,600 demonstrators were jailed, and on March 21 thousands of people from across the nation initiated a highly visible march from Selma to Montgomery. During the confrontations two Northern whites and a local Black demonstrator were murdered by Alabama lawmen. As in Birmingham, the Selma confrontations outraged the nation and drew widespread international attention. Once again the movement had maneuvered the Federal government into a corner where it had to play a role in resolving a crisis.

Because of the Birmingham movement and its aftermath, the Federal government issued national legislation outlawing all forms of racial segregation and discrimination. Thus, on June 2, 1964 President Johnson signed into law the 1964 Civil Rights Act. Similarly, to resolve the crisis created by Selma, the Congress passed national legislation that enfranchised Southern Blacks. Johnson signed the 1965 Voting Rights Bill into law on August 6, 1965. The 1964 Civil Rights Act and the 1965 Voting Rights Bill brought the formal regime of Jim Crow to a close.

By overthrowing Jim Crow in a matter of ten years, the civil rights movement had taught the nation and the world an important lesson: a mass-based grass roots social movement that is sufficiently organized, sustained, and disruptive is capable of generating fundamental social change. James Lawson captured it best when he concluded, "Many people, when they are suffering and they see their people suffering, they want direct participation. . . . So you put into the hands of all kinds of ordinary people a positive alternative to powerlessness and frustration. That's one of the great things about direct action" (Lawson 1978). Structural opportunities helped facilitate the rise of the modern civil rights movement, but it was the human agency of the Black community and its supporters that crushed Jim Crow.

NATIONAL AND INTERNATIONAL SIGNIFICANCE OF THE CIVIL RIGHTS MOVEMENT

Scholars of social movements have increasingly come to recognize the pivotal role that the civil rights movement has played in generating movements in America and abroad. A consensus is emerging that the civil rights movement was the catalyst behind the wave of social movements that crystallized in the United States beginning in the middle of the 1960s and continuing to the present (see Evans 1980, Freeman 1983, Morris 1984, 1993, Adam 1987, McAdam 1988, Snow & Benford 1992, Tarrow 1994, Groch 1998). This body of literature has shown that movements as diverse as the student movement, the women's movement, the farm workers' movement, the Native American movement, the gay and lesbian movement, the environmental movement, and the disability rights movement all drew important lessons and inspiration from the civil rights movement. It was the civil rights movement that provided the model and impetus for social movements that exploded on the American scene.

A myriad of factors were responsible for the civil rights movement's ability to influence such a wide variety of American social movements. The ideas that human oppression is not inevitable and that collective action can generate change were the most important lessons the civil rights movement provided other groups. As Angela Davis (1983) has pointed out, the Black freedom struggle in America has taught other groups about the nature of human oppression and important lessons about their own subjugation. This was certainly the case with the civil rights movement, and this message was amplified by the great visibility accorded the movement by mass media, especially television.

The legal achievements of the civil rights movement were crucial to its capacity to trigger other movements. The 1964 Civil Rights Act was paramount in this regard for it prohibited a wide array of discriminations based on race, color, religion, national origin, and sex (Whalen & Whalen 1985). The modem women's movement was first to take advantage of this aspect of the Civil Rights Act because women in Congress had to struggle to get the word "sex" included in the legislation precisely because its opponents recognized that the incorporation of sex would have far ranging implications for the place of women in society. The general point is that this legislation, and others generated by the civil rights movement, created the legal framework through which other groups gained the constitutional right to demand changes for their own population and they were to do so in the context of their movements for change.

A repertoire of collective action, as Tilly (1978) has pointed out, is crucial to the generation of protest. It was the civil rights movement that developed the repertoire utilized by numerous American social movements. This movement sparked the widespread use of the economic boycott, sit-ins, mass marches and numerous tactics that other movements appropriated. It also developed a cultural repertoire including freedom songs, mass meetings, and freedom schools that would also be utilized by a variety of movements. New movements could hit the ground running because the Black movement had already created the repertoires capable of fueling collective action.

The civil rights movement also revealed the central role that social movement organizations played in mobilizing and sustaining collective action. Organizations like SCLC, NAACP, CORE and SNCC were very visible in the civil rights movement and were the settings in which a variety of activists including women and students were drawn into civil rights activism. These social movement organizations provided the contexts that served as the training ground for activists who would return to their own populations to organize social movements.

Social movement theory has not developed a viable framework for understanding why at certain moments in history oppressed groups are able to develop the moral courage and make the extraordinary sacrifices that collective action requires. But the fact that African Americans were willing to be beaten, jailed, and killed for their activism while exuding unprecedented levels of dignity surely played an important role in encouraging other groups to confront authorities in their quest for change. As we glance backward, it becomes exceedingly clear that it was the civil rights movement that fertilized the ground in which numerous American social movements took root and flowered into widespread collective action.

The American civil rights movement has had an impact beyond the shores of America. Many of the same reasons the civil rights movement influenced American social movements, appear to also be the source of its influence on international movements. The major exception is that the civil rights movement did not serve as the training ground for many of the activists who initiated movements outside the United States. What is clear, however, is that numerous international movements were influenced by the US civil rights movement.

A similarity that movements share across the world is that they usually must confront authorities who have superior power. The major challenge for such movements is that they must develop a collective action strategy that will generate leverage enabling them to engage in power struggles with powerful opponents. The strategy of nonviolent direct action was first developed by Gandhi in South Africa and then used by Gandhi in the mass movement that overthrew British colonialism in India. Ghandi's use of nonviolence was important to the civil rights movement because some key leaders of the civil rights movement—James Farmer, Bayard Rustin, James Lawson and Glenn Smiley—had studied Gandhi's movement and became convinced that nonviolence could be used by African Americans. Additionally, Gandhi became a hero and a source of inspiration for Martin Luther King Jr. It was the American civil rights movement that perfected and modernized nonviolent direct action. Because of this achievement, the civil rights movement was the major vehicle through which nonviolent direct action was spread to other movements internationally. Nonviolent direct action has enabled oppressed groups as diverse as Black South Africans, Arabs of the Middle East, and pro-democracy demonstrators in China to engage in collective action. Leaders of these movements have acknowledged the valuable lessons they have learned from the civil rights movement (see Morris 1993). As Tarrow (1994) has pointed out, nonviolent direct action is a potent tool of collective action because it generates disruption and uncertainty that authorities must address. Tarrow captured how nonviolent direct action has spread domestically and internationally following the civil rights movement when he wrote:

Although it began as a tool of nationalist agitation in the Third World, nonviolent direct action spread to a variety of movements in the 1960s and 1970s. It was used in the Prague Spring, in the student movements of 1968, by the European and American Peace and environmental movements, by opponents of the Marcos regime in the Philippines and of military rule in Thailand and Burma (Tarrow 1994:109).

As McAdam (1995) has argued, it is possible that the lessons of the civil rights movement crossed national boundaries through complex diffusion processes. Although McAdam leaves the mechanisms of diffusion unspecified, interviews by the author with leaders of the movements in South Africa, China, and the West Bank provide insights. Mubarak Awad, one of the central leaders of the Intifadah Movement on the West Bank, conveyed that they absorbed the lessons of America's nonviolent direct action civil rights movement by acquiring films on the Black movement and Martin Luther King, and showing them on Jordanian television so that the lessons of that movement could be widely disseminated and applied (1990). Shen Tong, one of China's 1989 pro-democracy leaders revealed that "my first encounter with the concept of nonviolence was in high school when I read about Martin Luther King Jr. and Mahatma Gandhi" (1990). Patrick Lekota, a leader of the United Democratic Front in South Africa, explained that many of the leaders of the National African Congress gained their knowledge of the Black movement when they studied in the United States and that literature concerning the civil rights movement was widely spread in South Africa where "it was highly studied material" (1990). Similarly, in the early days of the Solidarity movement in Poland, Bayard Rustin, a major tactician of the civil rights movement, was summoned to Poland to give a series of colloquia and speeches on how nonviolent direct action worked in the civil rights movement. Summarizing the responses, Rustin stated, "I am struck by the complete attentiveness of the predominantly young audience, which sits patiently, awaiting the translation of my words" (Rustin, undated report). Additionally, leaders of these movements indicated that King, the charismatic leader who won the Nobel Peace Prize, was another important factor that fixed international attention upon the civil rights movement.

Because the civil rights movement developed a powerful tactical, ideological, and cultural repertoire of collective action available to a worldwide audience through mass media and an extensive literature, it has served as a model of collective action nationally and internationally. Awareness of the civil rights movement is so widespread globally that oppressed people in distant lands seek out knowledge of its lessons so they can employ it in their own struggles. Diffusion processes are important in this regard, but they merely complement the active pursuit of information pertaining to the civil rights movement by those wishing to engage in collective action here and abroad. The national anthem of the civil rights movement, "We Shall Overcome," continues to energize and strengthen the resolve of social movements worldwide.

THE CIVIL RIGHTS MOVEMENT
AND THE STUDY OF SOCIAL MOVEMENTS

In a recent book (1993), the sociologist James McKee explored the question as to why no sociological scholar anticipated the civil rights movement and Black protest that rocked the nation throughout the 1960s and early 1970s. After an extensive examination of the literature, he argued that this failure could not be attributed to the lack of theory or empirical data because the sociology of race relations had a wealth of both. Rather, he maintained that it was the assumptions underlying these theories that prevented sociologists from anticipating the civil rights movement. He argued that prior to the movement, sociologists viewed Southern rural Blacks as a culturally inferior, backward people. He concluded that "there was a logical extension of this image of the American Black: a people so culturally inferior would lack the capability to advance their own interests by rational action. . . . Blacks were portrayed as a people unable on their own to affect changes in race relations. . . ." (1993:8). In short, the study of race relations lacked an analysis of Black agency.

Collective behavior and related theories of social movements, which dominated the field prior to the 1970s, were similar in that they lacked a theory of Black agency. The problem was even more pronounced for theories of social movements prior to the civil rights movement, for those theories operated with a vague, weak version of agency to explain phenomena that are agency driven. Those theories conceptualized social movements as spontaneous, largely unstructured, and discontinuous with institutional and organizational behavior (Morris & Herring 1987). Movement participants were viewed as reacting to various forms of strain and doing so in a non-rational manner. In these frameworks, human agency was conceptualized as a reactive agency, created by uprooted individuals seeking to reestablish a modicum of personal and social stability. External uncontrollable factors were in the driver seat, directing the agency encompassed in collective action.

Resource mobilization and political process models of social movements reconceptualized social movement phenomena as well as the human agency that drives them. The civil rights movement and the movements it helped spur, were pivotal in the reconstruction of social movement theory. The civil rights movement provided a rich empirical base for the reexamination of social movements because the structures and dynamics of that movement could not be reconciled with existing social movement theories. Central themes of current social movement theory—the role of migration and urbanization, mobilizing structures including social networks, institutions and social movements organizations, tactical repertoires and innovations, dissemination of collective action, culture and belief systems, leadership, the gendering phenomenon and movement outcomes—have been elaborated in the context of the civil rights movement because those structures and processes were germane to that movement.

Theories prior to the civil rights movement had argued that migration and urbanization were vital to collective action because they produced the social strain and ruptured belief systems that drove people into collective action to reconstitute the social order. Scholars of the civil rights movement (Obserchall 1973, McAdam 1982, Morris 1984) have produced convincing evidence to the contrary. Black migration and urbanization facilitated the civil rights movement because it led to the institutional building and the proliferation of dense social networks across localities and across neighborhoods in cities through which the movement was mobilized and sustained. This finding helped solidify the proposition that migration and urbanization may facilitate social movements because of their capacity to produce and harden the social organization critical to the production of collective action.

The social organization underlying collective action encompasses a movement's mobilizing structures. These structures include formal and informal organizations, communication networks, local movement centers, social movement organizations and leadership structures. Morris (1984) and McAdam (1982) developed detailed analyses of the crucial role that Black churches, colleges and informal social networks played in the mobilization and development of the civil rights movement. Morris (1984) demonstrated that the civil rights movement was comprised of local movements and that it was local mobilizing structures that produced the power inherent in the civil rights movement. To capture this phenomenon, he developed the concept of "local movement center" and defined it as the social organization within local communities of a subordinate group, which mobilizes, organizes, and coordinates collective action aimed at attaining the common ends of that subordinate group. Without those mobilizing structures, it is doubtful that the movement would have been able to consolidate the resources required to confront and prevail over white racists and powerful state structures. These findings helped to discredit arguments maintaining that movements were spontaneous and discontinuous with pre-existing social structures.

The civil rights movement afforded scholars the opportunity to examine the fundamental role played by social movement organizations (SMOs) in the creation and coordination of collective action. That movement was loaded with SMOs operating at the local and national levels. By examining these SMOs, scholars (Morris 1984, Barkan 1986, Haines 1984) have assisted in the development of an interorganizational analysis of social movement organizations. Morris analyzed how each of the major SMOs of the civil rights movement shaped collective action by carving out its own spheres of organizational activity and producing the leaders, organizers, and tactics that provided the movement with its power and dynamism. At the interorganizational level, these organizations engaged in competition, cooperation, and conflict. What has been learned from this interorganizational standpoint, is that when SMOs compete and cooperate they can produce greater volumes of collective action by sharing knowledge, and resources and by triggering tactical innovations. SMOs can be destructive to social movements when they engage in intense conflict and generate warring factions. This appears to have been the case in Albany, Georgia in 1962 when the movement failed to reach its goals because of the conflict between SCLC and SNCC (see Morris 1984:239–250). By examining the civil rights movement, Haines (1984) demonstrated how interorganizational relations between SMOs can affect the outcomes of a movement because the agenda of radical organizations can cause authorities, because of their fear of radical alternatives, to concede to the demands of moderate SMOs. Research prior to the civil rights movement tended to conceive SMOs as by-products of social movements and thus as not germane to their causation and outcomes. The visibility of SMOs in the civil rights movement and their obvious centrality, have helped scholars to reconceptualize the role of SMOs and give them the theoretical attention they merit.

The civil rights movement has served as a rich empirical base for the analysis of social movement tactics. The widespread use and development of nonviolent direct action tactics is one of the crowning achievements of the civil rights movement. The labor movement and other grassroots movements of the 1930s adopted versions of some of the tactics—the sit-down strike, freedom songs, labor schools, community-based organizing—that would be used by the civil rights movement. Nevertheless, it was the civil rights movement that expanded these tactics and employed them in contexts where they would have enormous impact world-wide. Indeed, the importance of the tools of nonviolent direct action and especially the sit-in, which the civil rights movement utilized and made famous around the globe, led Tarrow to declare that these tactics were "perhaps the major contributions of our century to the repertoire of collective action" (Tarrow 1994:108). Several formulations by scholars examining the tactical repertoire of the civil rights movement have been developed. In an analysis of the 1960 sit-ins, Morris (1981) demonstrated that this tactic spread rapidly because of the mobilizing structures already in place by 1960, and because the tactics of nonviolent direct action fitted into the ideological and organizational framework of the Black church, which preached against violence and extolled the virtues of redemptive suffering.

In an analysis of tactical innovations within the civil rights movement, McAdam (1983) examined how tactical innovations increased the collection action of that movement and affected its outcomes. He argued that once authorities learn to neutralize a tactic, movement leaders had to devise a new tactic or risk a decline in collection action and the defeat of the movement. A study by Morris (1993) confirmed the important role of tactical innovation but took issue with McAdam as to how the process worked in the civil rights movement. He found that the tactical innovation process of the civil rights movement consisted of activists developing and adding new tactics to existing ones and employing them dynamically. The use of multiple tactics increased the possibility of generating a crisis, which served as the leverage to achieve movement demands. A great deal more theoretical work on movement tactics is needed and the civil rights movement will continue to provide a rich terrain for such theorizing.

The development of cycles of protest is intimately related to the development of tactical and cultural repertoires. Tarrow (1994:154) has defined a cycle of protest as "a phase of heightened conflict and contention across the social system." During such phases, collective action diffuses rapidly from more mobilized to less mobilized sectors and it generates "sequences of intensified interactions between challengers and authorities" (p. 154). A cycle of protest occurred in the United States following the civil rights movement when numerous movements—antiwar, women, farm workers, Native Americans, gay and lesbians, etc.—sprang into action. This cycle was not limited to national boundaries; it was also replicated in Europe during the same period. A major theoretical task is to explain the cycles of protest (McAdam 1995). The civil rights movement has figured prominently in such theorizing because it was the initiator movement that set this cycle in motion. Tarrow (1994) and McAdam (1995) have utilized the case of the civil rights movement to argue that such cycles are likely to occur when there exists an initial movement that has been successful in developing a repertoire of tactical, organizational and ideological lessons that can be transplanted to other movements under conditions favorable to social protest. Because the civil rights movement provided such a clear example of a protest cycle initiator, it continues to inform scholars about the processes by which a "family of movements" burst on the social scene.

CULTURE AND CIVIL RIGHTS MOVEMENT

Social movement scholars are increasingly coming to recognize that cultural factors weigh heavily in collective action (see Morris & McClurg Mueller 1992). In order for people to be attracted to and participate in collective action, they must come to define a situation as intolerable and changeable through collective action. To do so they develop injustice frames (Gamson 1992) and undergo cognitive liberation (McAdam 1982). Black cultural beliefs and practices were pronounced in the civil rights movement, and they affected strategic choices in that movement and figured significantly in the cycles of protest that it helped trigger. Morris (1984) analyzed how Black music, prayers, and religious doctrines were refashioned to critique Black oppression and to promote solidarity and to function as a tool of mobilization for the civil rights movement.

David Snow and his colleagues (1986, 1992) have produced the most compelling formulation as to why belief systems are germane to collective action. Parts of their analysis rest squarely on dynamics of the civil rights movement. Snow and his colleagues (1986) argued that framing processes constitute one of the major avenues through which collective action is generated, disseminated, and sustained. For them, frame alignment is the key process because it refers to "the linkage of individual and SMO interpretive orientations, such that some set of individual interests, values and beliefs and SMO activities, goals and ideology are congruent and complementary" (1986:484). Thus, SMO leaders are able to recruit and mobilize movement participants through creative ideological work whereby frames are bridged, amplified and extended. Snow & Benford's (1992) concept of an innovative master frame is explicitly derived from an analysis of the civil rights movement. Master frames are generic frames of meaning and one of their variable features is their capacity to generate "diagnostic attribution, which involved the identification of a problem and the attribution of blame or causality" (Snow & Benford 1992: 138). When an innovative master frame becomes elaborated it "allows for numerous aggrieved groups to tap it and elaborate their grievances in terms of its basic problem-solving schema" (p. 140). Snow & Benford argued that the civil rights movement developed an elaborate civil rights master frame. That master frame was so significant because its "punctuation and accentuation of the idea of equal rights and opportunities amplified a fundamental American value that resonated with diverse elements of Americans society and thus lent itself to extensive elaboration" (p. 148). Because of its master frame, they argued that the civil rights movement was able to play a pivotal role in generating cycles of protest. More theorizing is needed on the central role that culture plays in social movements, and the civil rights movement will continue to be a key reservoir for such work because much of what it has to teach scholars about culture and collective action remains untapped.

GENDER AND CIVIL RIGHTS MOVEMENT

Rosa Parks is known as the mother of the civil rights movement because her refusal to give a white man her seat on a segregated bus sparked the Montgomery, Alabama bus boycott. Parks' resistance and civil rights activity were not unlike those of countless Black women such as Ella Baker, Septima Clark, Fannie Lou Hammer, and Diane Nash (see Crawford et al. 1990, Barnett 1993, Payne 1994, Robnett 1997). Black women played crucial roles in the civil rights but they did so in the context of a society and a movement characterized by high levels of patriarchy. Morris (1984) detailed the roles that Black women played in the origins of the civil rights movement, and he discussed how their contributions were restricted by male sexism within the movement.

Recent studies (Barnett 1993, Payne 1995, Robnett 1997) examining the role of women in the civil rights movement have sought to understand how the movement itself was gendered and how this reality affected movement dynamics. Payne and Barnett analyzed how Black women played pivotal organizing roles, often behind the scene, that enabled the movement to perform the multifaceted, mobilization and organizational tasks crucial for wide-scale collective action. These studies, however, make it clear that these women performed these duties not because of natural inclinations to work behind the scene. They did so because they were dedicated to the goals of the movement and because male-dominated hierarchies limited the types of contributions they could make.

Robnett's study (1997) has produced some important insights on how gendering affected the civil rights movement. She found that despite the sexism, the contributions of many Black women encompassed a great deal more than backstage organizing. Black women constituted the bridge leaders of the civil rights movement. Bridge leadership is "an intermediate layer of leadership, whose tasks include bridging potential constituents and adherents, as well as potential formal leaders, to the movement" (Robnett 1997:191). Because these women were excluded from formal leadership positions because of their gender, they could act in more radical ways than men, given their allegiances to the grassroots and their freedom from state constraints. Given the structural location of this leadership it had more latitude to do the movement's emotional work which increased its mobilization capacity and generated greater strategic effectiveness. The majority of bridge leaders in the civil rights movement were Black women because that "was the primary level of leadership available to women" (Robnett 1997:191). Scholars such as Robnett have just begun to unpack the gendering dynamic of movements. The civil rights movement provides a rich empirical case for the continuance of this work because it was a movement in which women and men converged in their activity for the sake of liberating a people.

LEADERSHIP AND CIVIL RIGHTS MOVEMENT

Two theoretical issues thrust up by the civil rights movement have not received much attention. We do not have a theory that explains the relationship between preexisting protest traditions and the rise and trajectory of new social movements. Black leaders from previous movements played important roles in the civil rights movement. The expertise of leaders such as Ella Baker, A Philip Randolph, Bayard Rustin, and Septima Clark supplied the movement with vision and technical knowledge pertaining to collective action. By the time of the modem civil rights movement, the NAACP had developed a sophisticated legal strategy to attack racial segregation. This strategy became one of the central features of the nonviolent direct action movement. A preexisting Black culture containing a variety of oppositional themes supplied the modem movement with much of its cultural content and anchored it to the cultural traditions of the Black community. There can be little doubt that this preexisting protest tradition help to solidify this movement and affected its trajectory from the outset. A theoretical formulation of preexisting protest traditions is needed for a comprehensive understanding of movement emergence and outcomes.

Current social movement theory has made little progress in the analysis of social movement leadership. Movement scholars have too readily assumed that movement leadership is a matter of common sense not requiring theoretical analysis (see McAdam et al. 1988:716). The civil rights movement reveals, however, that movement leadership is a complex phenomenon that remains relatively unexplored theoretically. Morris (1984) found that charismatic leadership played a key role in the civil rights movement but functioned differently from the way it was characterized in Weber's classic formulation, given that it was rooted in preexisting institutions and organizational structures. Robnett's (1997) finding concerning bridge leadership reveals the complex composition that leadership structures can assume in movements and the way such structures are shaped by durable practices in the larger society. Additional theorizing is needed to uncover the processes by which individuals are chosen to become movement leaders and how factors internal to the movement constrain the strategic options available to them. The social class of leaders appears to be important in determining leadership style and the degree to which they can be successful in mobilizing movement participants across social classes. Thus, as the civil movement reveals, movement leadership is a complex variable phenomenon that should receive attention from social movement scholars.

LOOKING BACKWARD TO MOVE FORWARD

By looking back we can appreciate the tremendous contributions the civil rights movement has made to social change and to the reconceptualization of an intellectual discipline. Because of that movement and those it triggered, social movement scholars have had to formulate new ideas and rethink why and how social movements continue to reshape the social landscape. As Gamson put it:

> If there hadn't been a civil rights movement there might not have been an anti-war movement, if there hadn't been these movements there might not have been an environmental movement. Without these movements there wouldn't have been people coming into the field who were receptive to a new orientation (Morris & Herring 1987:184).

The civil rights movement helped trigger a paradigmatic shift in the field of social movements and collective action. It did not do so alone, for other historic and contemporary movements worldwide have provided empirical puzzles that have assisted in the reconceptualization of the field. But the intellectual work based on the civil rights movement has been substantial and transformational. This movement has provided scholars with empirical and theoretical puzzles that should continue to push the field forward until new path—breaking social movements arise and shake up the field once again.

I close by returning to McKee's issue regarding Black agency. The civil rights movement revealed that that agency resides in Black social networks, institutions, organizations, cultures, and leaders, and in the creativity of a people who have had to engage in a chain of struggles to survive relentless oppression and to maintain their dignity. What is more, that movement helped other groups to locate their own agency, who then harnessed it in activities that have generated important changes in America and the world.

ACKNOWLEDGMENTS

I would like to thank professors Jeff Manza, Arthur Stinchcombe, and Mayer Zald for their insightful comments on an earlier version of this chapter. I also thank Kiana Morris for her assistance in the preparation of the manuscript.

LITERATURE CITED

Adam B. 1987. *The Rise of a Gay and Lesbian Movement*. Boston: Twayne

Awad M. 1990. *Interview*, Feb. 9, Boston, Mass

Barkan S. 1986. Interorganizational Conflict in the Southern Civil Rights Movement. *Social. In.* 56:190–209

Barnett B. 1993. Invisible black women leaders in the civil rights movement: the triple constraints of gender, race and class. *Signs* 7(2):162–82

Bloom JM. 1987. *Class, Race and the Civil Rights Movement*. Bloomington: Indiana Univ. Press

Blumberg RL. 1984. *Civil Rights: The 1960s Freedom Struggle*. Boston: Twayne

Bobo L. 1997. The color line, the Dilemma, and the dream: race relations in America at the close of the twentieth century. In *Civil Rights and Social Wrongs: Black-White Relations Since World War II*, ed. J Higham, pp. 31–55. University Park, Penn: Penn. State Univ. Press

Branch T. 1988. *Parting the Waters: America in the King Years*. New York: Simon & Schuster

Carson C. 1981. *In Struggle: SNCC and the Black Awakening of the 1960s*. Cambridge: Harvard Univ. Press

Collier-Thomas B. 1984. *Black Women Organized for Social Change 1800–1920*. Washington DC: Bethune Mus. Arch.

Crawford V, Rouse J, Woods B. 1990. *Women in the Civil Rights Movement*. Brooklyn, NY: Carlson

Davis A. 1981. *Women, Race and Class*. New York: Random House

DuBois WEB. 1903. *The Souls of Black Folk*. Greenwich, CT: Fawcett

Evans S. 1980. *Personal Politics*. New York: Vintage Books

Fairclough A. 1987. *To Redeem the Soul of America: The Southern Christian Leadership Conference and Martin Luther King, Jr.* Athens: The Univ. Georgia Press

Franklin JH. 1967. *From Slavery to Freedom: A History of American Negroes*. New York: Knopf

Fredrickson GM. 1995. *Black Liberation*. New York: Oxford Univ. Press

Freeman J, ed. 1983. *Social Movements of the Sixties and Seventies.* New York: Longman

Gamson W. 1992. The social psychology of collective action. See Morris & McClurg Mueller 1992, pp. 53–76

Garfinkel H. 1969. *When Negroes March: The March on Washington Movement in the Organizational Politics for FEPC.* New York: Atheneum

Garrow D. 1986. *Bearing the Cross: Martin Luther King, Jr., and the Southern Christian Leadership Conference.* New York: Morrow & Co

Groch S. 1998. Pathways to protest: the making of oppositional consciousness by people with disabilities. PhD thesis. Northwestern Univ., Evanston.

Haines H. 1984. Black radicalization and the funding of civil rights: 1957–1970. *Soc. Prob. 32*:31–43

Harding V. 1983. *There Is a River: The Black Struggle for Freedom in America.* New York: Vantage Books

Jaynes G, Williams R Jr, eds. 1989. *A Common Destiny: Blacks and American Society.* Washington, DC: Natl. Acad. Press

Kluger R. 1975. *Simple Justice.* New York: Knopf

Ladner J. 1979. The South: Old-New Land. *New York Times*, May 17, 1973, p. 23

Lawson J. 1978. Interview, October 2. Los Angeles, Calif

Lawson S. 1991. *Running for Freedom.* Philadelphia: Temple Univ. Press

Lekota P. 1990. Interview. February 9, Boston, Mass

Mays BE. 1978. Interview. September 20. Atlanta, Ga

McAdam D. 1982. *Political Process and the Development of Black Insurgency.* Chicago: Univ. Chicago Press

McAdam D. 1983. Tactical innovation and the pace of insurgency. *Am. Social Rev. 48*:735–54.

McAdam D. 1988. *Freedom Summer.* New York: Oxford Univ. Press.

McAdam D. 1995. "Initiator" and "Spin-off" movements: diffusion processes in protest cycles. In *Repertoires and Cycles of Collective Action,* ed. M Traugott, pp. 217–39. Durham, NC: Duke Univ. Press.

McAdam D, McCarthy J, Zald M. 1988. Social movements. In *Handbook of Sociology,* ed. NJ Smelser, p. 695–737. Newbury Park, CA: Sage.

McKay C. 1963. If we must die. In *American Negro Poetry,* ed. A Bontemps, p. 31. New York: Hill & Wang

McKee JB. 1993. *Sociology and the Race Problem.* Urbana: Univ. Ill. Press

McNeil G. 1983. *Groundwork: Charles Hamilton Houston and the Struggle for Civil Rights.* Philadelphia: Univ. Penn. Press

Meier A, Rudwick E. 1976. *Along the Color Line. Explorations in the Black Experience.* Urbana: Univ. Ill. Press

Moody A. 1968. *The Coming of Age in Mississippi.* New York: Dial Press

Morris A. 1981. Black southern student sit-in movement: an analysis of internal organization. Am. Social. Rev. 46:744–67

Morris A. 1984. *The Origins of the Civil Rights Movement: Black Communities Organizing for Change.* New York: Free Press

Morris A. 1993. Birmingham confrontation reconsidered: an analysis of the dynamics and tactics of mobilization. *Am. Social. Rev.* 58:621–36

Morris A, Herring C. 1987. Theory and research in social movements: a critical review, *Annu. Rev. Polit. Sci.* 2:137–98

Morris A, McClurg Mueller C, eds. 1992. *Frontiers in Social Movement Theory.* New Haven, CT: Yale Univ. Press

Oberschall A. 1973. *Social Conflict and Social Movements.* Englewood Cliffs, NJ: Prentice-Hall

Payne C. 1994. *I've Got the Light of Freedom.* Berkeley: Univ. Calif. Press

Robnett B. 1997. *How Long? How Long?* New York: Oxford Univ. Press

Rustin B. No date. *Report on Poland.* New York: A. Philip Randolph Inst.

Sale K. 1973. *SDS.* New York: Vintage Books

Schlesinger AM Jr. 1965. *A Thousand Days.* Boston: Houghton Mifflin

Sitkoff H. 1981. *The Struggle for Black Equality, 1945–1980.* New York: Hill & Wang

Snow D, Benford R. 1992. *Master frames and cycles of protest.* See Morris & McClurg Mueller 1992, pp. 133–55

Snow D, Rochford E Jr, Worden S, Benford R. 1986. Frame alignment processes, micro-mobilization, and movement participation. *Am. Soc. Rev.* 51:464–81

Sterling C, Kittross J. 1978. *Stay Tuned: A Concise History of American Broadcasting.* Belmont: Wadsworth

Tarrow S. 1994. *Power in Movement: Social Movements, Collective Action and Politics.* Cambridge Univ. Press

Tilly C. 1978. *From Mobilization to Revolution.* Reading, Mass: Addison-Wesley

Tong S. 1990. *Shen Tong's King Center Address.* Atlanta, GA: Martin Luther King Cent. Soc. Change

Tushnet M. 1987. *The NAACP'S Legal Strategy against Segregated Education, 1925–1950.* Chapel Hill: Univ. NC Press

Whalen C, Whalen B. 1985. *The Longest Debate: A Legislative History of the 1964 Civil Rights Act.* Cabin John, MD.: Seven Lakes Press

Whitfield S. 1988. *A Death In the Delta: The Story of Emmett Till.* New York: Free Press

InfoMarks: Make Your Mark

What Is an InfoMark?

It's a single-click return ticket to any page, any result, any search from InfoTrac College Edition.

An InfoMark is a stable URL, linked to InfoTrac College Edition articles that you have selected. InfoMarks can be used like any other URL, but they're better because they're stable—they don't change. Using an InfoMark is like performing the search again whenever you follow the link—whether the result is a single article or a list of articles.

How Do InfoMarks Work?

If you can "copy and paste," you can use InfoMarks.

When you see the InfoMark icon on a result page, its URL can be copied and pasted into your electronic document—Web page, word processing document, or email. Once InfoMarks are incorporated into a document, the results are persistent (the URLs will not change) and are dynamic.

Even though the saved search is used at different times by different users, an InfoMark always functions like a brand new search. Each time a saved search is executed, it accesses the latest updated information. That means subsequent InfoMark searches might yield additional or more up-to-date information than the original search with less time and effort.

Capabilities

InfoMarks are the perfect technology tool for creating:

- Virtual online readers
- Current awareness topic sites—links to periodical or newspaper sources
- Online/distance learning courses
- Bibliographies, reference lists
- Electronic journals and periodical directories
- Student assignments
- Hot topics

Advantages

- Select from over 15 million articles from more than 5,000 journals and periodicals
- Update article and search lists easily
- Articles are always full-text and include bibliographic information
- All articles can be viewed online, printed, or emailed
- Saves professors and students time
- Anyone with access to InfoTrac College Edition can use it
- No other online library database offers this functionality
- FREE!

How to Use InfoMarks

There are three ways to utilize InfoMarks—in HTML documents, Word documents, and Email

HTML Document

1. Open a new document in your HTML editor (Netscape Composer or FrontPage Express).
2. Open a new browser window and conduct your search in InfoTrac College Edition.
3. Highlight the URL of the results page or article that you would like to InfoMark.
4. Right click the URL and click Copy. Now, switch back to your HTML document.
5. In your document, type in text that describes the InfoMarked item.
6. Highlight the text and click on Insert, then on Link in the upper bar menu.
7. Click in the link box, then press the "Ctrl" and "V" keys simultaneously and click OK. This will paste the URL in the box.
8. Save your document.

Word Document

1. Open a new Word document.
2. Open a new browser window and conduct your search in InfoTrac College Edition.
3. Check items you want to add to your Marked List.
4. Click on Mark List on the right menu bar.
5. Highlight the URL, right click on it, and click Copy. Now, switch back to your Word document.
6. In your document, type in text that describes the InfoMarked item.

7. Highlight the text. Go to the upper bar menu and click on Insert, then on Hyperlink.
8. Click in the hyperlink box, then press the "Ctrl" and "V" keys simultaneously and click OK. This will paste the URL in the box.
9. Save your document.

Email

1. Open a new email window.
2. Open a new browser window and conduct your search in InfoTrac College Edition.
3. Highlight the URL of the results page or article that you would like to InfoMark.
4. Right click the URL and click Copy. Now, switch back to your email window.
5. In the email window, press the "Ctrl" and "V" keys simultaneously. This will paste the URL into your email.
6. Send the email to the recipient. By clicking on the URL, he or she will be able to view the InfoMark.